T0134562

Mathematical Modelling of Haemodialysis

Leszek Pstras • Jacek Waniewski

Mathematical Modelling of Haemodialysis

Cardiovascular Response, Body Fluid Shifts, and Solute Kinetics

Leszek Pstras
Polish Academy of Sciences
Nalecz Institute of Biocybernetics and
Biomedical Engineering
Warsaw, Poland

Jacek Waniewski
Polish Academy Science
Nalecz Institute of Biocybernetics and
Biomedical Engineering
Warsaw, Poland

ISBN 978-3-030-21412-8 ISBN 978-3-030-21410-4 (eBook)
https://doi.org/10.1007/978-3-030-21410-4

This Springer imprint is published by the registered company Springer Nature Switzerland AG
The registered company address is: Gewerbestrasse 11, 6330 Cham, Switzerland

Preface

Almost four million people around the world are diagnosed with end-stage renal disease (ESRD), i.e. a chronic kidney failure. Most of these patients (over 80%) receive regular dialysis treatment to substitute the most important, life-saving functions of a healthy kidney, i.e. to remove excess water, waste products of metabolism and foreign chemicals from the body, to maintain the physiological levels of electrolytes and to correct the metabolic acidosis. The most common dialysis treatment is haemodialysis (HD) – an extracorporeal technique – in which the patient's blood is cleared from unwanted solutes and excess water in a dialyzer.

While being a life-sustaining treatment for patients with ESRD, haemodialysis imposes a significant burden on the cardiovascular system. During a standard HD session, a few litres of fluid (water and solutes) are removed from the body over a relatively short period of time (typically 3–5 hours). The majority of dialysed patients can deal with the reduction in blood volume and have their blood pressure modestly reduced or unchanged. For some patients, however, the relatively rapid changes in the fluid status during HD can cause several problems, such as intradialytic hypotension or hypertension, the pathogenesis of which is still incompletely understood.

In light of the continually growing number of dialysis patients worldwide and the presence of the aforementioned clinical problems associated with haemodialysis therapy, the main aim of this publication was to develop and present a comprehensive mathematical model that integrates cardiovascular dynamics with the whole-body water and solute kinetics to simulate and analyse the transport and regulatory processes taking place in the patient's body during HD. In particular, the aim was to build on, combine, improve and extend the existing models in order to develop a physiologically based compartmental model of transport of ions (sodium, potassium, chloride, bicarbonate and other cations or anions), small electroneutral molecules (urea and creatinine), proteins (albumin and globulins), red blood cells, plasma and water within the body and across the dialyzer to analyse changes in blood pressure, blood volume and other haemodynamic variables in patients undergoing HD.

Beginning with an introduction to kidney function and renal replacement therapies, the book explores the physiological principles of the short-term cardiovascular baroreflex regulation and the mechanisms of water and solute transport across the human body providing a detailed description of the proposed model structure, model equations and associated assumptions. The model-based simulations are then used to analyse the baroreflex mechanisms, osmotic water shifts and solute transport during HD and to discuss the mechanisms potentially involved in the pathogenesis of intradialytic hypotension or hypertension.

Including an up-to-date review of the literature related to the modelled physiological mechanisms and processes, the book should serve both as an overview of transport and regulatory mechanisms related to the cardiovascular system and body fluids and as a comprehensive example of employing mathematical models to address clinically important issues.

The book is intended for the researchers and graduate students in biomedical engineering, physiology or medicine interested in mathematical modelling of cardiovascular dynamics and water and solute transport across the human body, both under physiological conditions and during haemodialysis therapy.

Warsaw, Poland Leszek Pstras
March, 2019 Jacek Waniewski

Acknowledgements

We would like to thank Professor Wojciech Załuska and Dr. Alicja Wójcik-Załuska from the Medical University of Lublin in Poland for sharing the results of their clinical studies on dialysis patients used for the validation of the model presented in this book.

We are also grateful to our colleagues from the Laboratory of Mathematical Modelling of Physiological Processes at the Nalecz Institute of Biocybernetics and Biomedical Engineering, Polish Academy of Sciences, for their help and support. In particular, we would like to thank Dr. Małgorzata Dębowska for her valuable help with the processing and interpretation of clinical data from dialysis patients, Dr. Jan Poleszczuk for sharing his experience in using mathematical and computer tools for modelling biological and physiological phenomena, Dr. Mauro Pietribiasi for the valuable discussions on mathematical modelling and statistics and Dr. Joanna Stachowska-Piętka for her input related to the microvascular transport processes.

The work presented in this book was supported by the doctoral research project 'Mathematical modelling of cardiovascular response to haemodialysis' funded by the Polish National Science Centre (agreement no. UMO-2016/20/T/ST7/00289).

Contents

Nomenclature

Abbreviations

CVS	Cardiovascular system
HCT	Haematocrit [%]
HD	Haemodialysis
HGB	Haemoglobin
HR	Heart rate [bpm]
MAP	Mean arterial pressure (mean pressure in large arteries) [mm Hg]
MCV	Mean corpuscular volume of red blood cells [fL]
MW	Molecular weight [g/mol or Da]
PV	Plasma volume
RBC	Red blood cells
SBP	Systolic blood pressure [mm Hg]
SV	Stroke volume [mL]
TBP	Total blood proteins [g/dL]
TBV	Total blood volume [L]
TBW	Total body water [L]
TPP	Total plasma proteins [g/dL]
TP	Total proteins [g/dL]
URR	Urea reduction rate [%]
VM	Valsalva manoeuvre

Greek Symbols

α	Gibbs-Donnan coefficient [-]
β	Equilibrium concentration ratio across the cellular membrane [-]
η	Blood dynamic viscosity [Pa · s]
κ	Parameter related to vascular resistance [mm Hg · s · mL]
π	Oncotic pressure [mm Hg]

ρ	Protein density [g/cm^3]
σ	Reflection coefficient [-]
τ	Baroreflex time constant [s]
φ	Osmotic activity coefficient [-]

Symbols

A	Red blood cell surface area [m^2]
c	Solute concentration [mmol/L or g/dL]
C	Compartment compliance [mL/mmHg]
d	Diameter of red blood cells [µm]
D	Diffusive dialysance [mL/min]
E	Heart contractility [-]
f	Parameter describing mean transmembrane solute concentration [-]
F	Water fraction [-]
g	Solute generation [mmol/24 h]
G	Baroreflex gain [1/mmHg; s/mmHg; mL/mmHg; s/mL]
J	Water flow rate [mL/s]
k	Slope of the baroreflex sigmoidal function [s; mL; mmHg.s/mL]
K	Transcellular mass transport coefficient [L/min]
$L\mathrm{p}$	Hydraulic conductivity of capillary walls [mL/min/mmHg]
LS	Lymph flow sensitivity to interstitial pressure changes [mL/mmHg/h]
M	Mass [g]
n	Number of red blood cells [-]
N	Number of moles of a given solute [mol]
O	Osmolarity [mOsm/L]
p	Solute permeability [cm/s]
P	Hydrostatic/hydraulic pressure [mm Hg]
Pe	Peclet number [-]
PS	Permeability-surface product [mL/h]
q	Cardiac output [L/min]
Q	Flow rate [mL/s]
Q_s	Molar transfer of solute s [mmol/s]
Q_uf	Dialyzer ultrafiltration [mL/min]
R	Vascular resistance [mmHg · s / mL]
S	Surface [m^2]
t	Thickness of red blood cells [µm]
T	Heart period [Hz]
V	Compartment volume [mL]
V_u	Compartment unstressed volume [mL]
w	Weighing factor [-]
z	Solute charge [-]

Subscripts

a	Arterial
ac	Arteriovenous access
at	Arterial tubing
b	Blood
c	Cardiopulmonary
C	Central (haematocrit)
cap	Capillary wall
cell	Cellular membrane
Cr	Creatinine
d	Dialyzer
D	Discharge (haematocrit)
ex	Excluded (volume)
ic	Intracellular
in	Inflow
is	Interstitial
l	Left
la	Large arteries
lat	Left atrium
lh	Left heart
lv	Large veins
n	Normal
out	Outflow
pa	Pulmonary arteries
pl	Blood plasma
pv	Pulmonary veins
r	Right
rat	Right atrium
rc	Red blood cells
rh	Right heart
s	Solute
sa	Small arteries
sc	Systemic capillaries
sv	Small veins
th	Intrathoracic
T	Tube (haematocrit)
U	Urea
vt	Venous tubing
w	Water

About the Authors

Leszek Pstras is an Assistant Professor at the Laboratory of Mathematical Modelling of Physiological Processes at the Nalecz Institute of Biocybernetics and Biomedical Engineering, Polish Academy of Sciences. His research focuses on mathematical modelling of cardiovascular short-term regulatory mechanisms as well as fluid and solute transport kinetics during haemodialysis therapy.

Jacek Waniewski is a Professor of Biomedical Engineering and the Chairman of the Scientific Council at the Nalecz Institute of Biocybernetics and Biomedical Engineering, Polish Academy of Sciences. His research focuses on mathematical modelling of membrane transport in medicine and biomedical engineering with a special interest in modelling transport processes during renal replacement therapies (both haemodialysis and peritoneal dialysis).

Chapter 1
Introduction to Renal Replacement Therapy

Abstract This chapter provides a background on kidney function, chronic kidney failure and renal replacement therapies with a particular emphasis on haemodialysis and the clinical problems associated with haemodialysis treatment, such as intradialytic hypotension or hypertension. It also provides an introduction to compartmental modelling of whole-body water and solute transport for modelling haemodialysis therapy as well as an overview of mathematical models of cardiovascular system and baroreflex regulation.

Keywords Kidney function · Chronic kidney failure · Renal replacement therapy · Haemodialysis · Blood pressure · Intradialytic hypotension · Intradialytic hypertension · Disequilibrium syndrome · Mathematical modelling · Compartmental model · Lumped-parameter model · Baroreflex · Fluid transport · Solute kinetics · Dialysis adequacy

1.1 Kidney Function

The kidneys are the major organs maintaining homeostasis of the human body. Their two essential functions are to regulate the volume and composition of body fluids and to remove from the body redundant or toxic substances, which are either produced in the body during metabolism or get to the body from the environment via oral, pulmonary or other routes.

In a typical middle-age male human, almost 60% of the total body weight is composed of water (the value for females is slightly lower – around 50%) [1], which is distributed among all body cells (intracellular fluid) and the fluids outside the cells (extracellular fluid), such as interstitial fluid, blood plasma or transcellular fluids (e.g. peritoneal, pleural, pericardial or cerebrospinal fluids). Within all these fluid spaces, a large number of solutes (ions or molecules, electrolytes or nonelectrolytes), such as minerals, proteins, carbohydrates, lipids, gases, vitamins, enzymes, hormones and wastes, are dissolved in water.

L. Pstras, J. Waniewski, *Mathematical Modelling of Haemodialysis*,
https://doi.org/10.1007/978-3-030-21410-4_1

In order to maintain a relatively constant volume and a stable composition of the body fluids, the kidneys continually adjust the excretion or reabsorption rate of water and electrolytes (sodium, potassium, chloride, etc.) to match the rate of their intake in the form of liquids or food and the synthesis of water through carbohydrates oxidation, taking into account the fluid lost in sweat and faeces as well as the insensible water loss through the respiratory tract and diffusion through the skin [2].

The kidneys eliminate from the body waste products of metabolism, such as urea (from plasma proteins catabolism), creatinine (from muscle proteins catabolism), uric acid (from nucleic acids catabolism), bilirubin (from haemoglobin catabolism) and metabolites of various hormones [2]. They are also responsible for expelling in the urine foreign chemicals, such as drugs, toxins, pesticides, food additives or other harmful substances that get to the body from the environment [2, 3].

Apart from maintaining the body's electrolyte-water balance and removing waste or toxic materials, the kidneys perform a variety of other vital tasks [2, 3] (see Fig. 1.1) including:

- Keeping a stable pH within the body by excreting acids (such as sulfuric, phosphoric or lactic acids) and replenishing the buffer system
- Regulating the arterial blood pressure (BP) through the renin-angiotensin-aldosterone system in the short term and through water and sodium excretion in the long term
- Secreting the hormone erythropoietin to stimulate the production of red blood cells
- Producing the active form of vitamin D – calcitriol (essential for promoting calcium deposition in bones)
- Synthesising glucose during prolonged fasting (gluconeogenesis)

In a normal healthy human, around 1500 L of blood flow through the kidneys each day, and hence the whole blood is cleansed circa 300 times a day. The healthy kidneys produce around 1.5 L of urine per day, although the rate of urine excretion

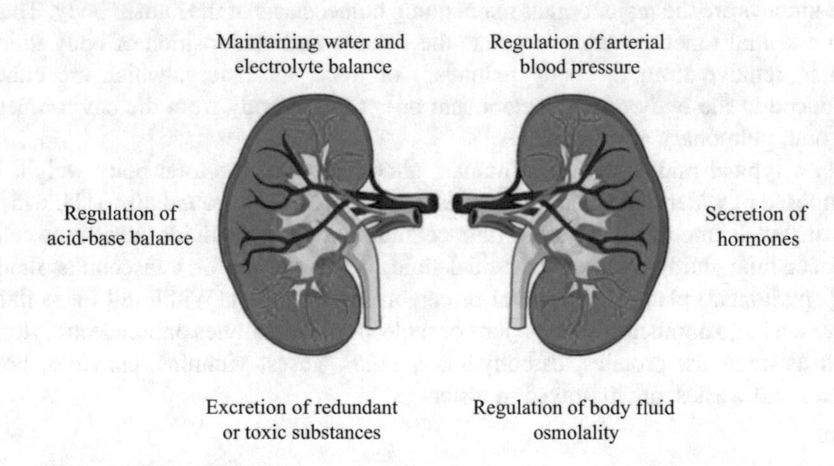

Fig. 1.1 Overview of kidney functions

depends strongly on the intake of water and can be as low as 0.5 L/day in a dehydrated person or as high as 20 L/day in a person drinking a lot of fluids [2].

1.2 Chronic Kidney Failure

Chronic kidney disease (CKD) consists in a gradual reduction of the functioning kidney tissue and the loss of kidney function. The CKD Work Group of the US National Kidney Foundation defines CKD as abnormalities of kidney structure or function, present for over 3 months, with implications for health [4]. CKD may not provide symptoms until the kidney function is significantly impaired or when irreversible damage to kidneys has occurred. The symptoms of an apparent CKD may include nausea, loss of appetite, fatigue, weakness, sleep problems, reduced mental sharpness, muscle cramps, swelling of ankles or itching [5, 6]. The last stage of a progressing CKD, i.e. stage 5, when the glomerular filtration rate (GFR) falls below 15 mL/min/1.73 m^2 [4], is known as chronic renal failure (CRF) or end-stage renal disease (ESRD) and describes the inability of the kidneys to effectively remove excess fluid from the body and to filter out waste products from the blood. The number of patients suffering from CRF worldwide reached almost 4 million in 2017 (see Table 1.1) [7].

Diseases such as diabetes, hypertension, chronic nephritis or polycystic kidney disease, as well as excessive intake of certain medications, can lead to CKD and eventually to CRF [5, 6]. Factors that may increase the risk of CKD include obesity, smoking, family history of kidney disease or older age [5, 8].

When the kidneys are irreparably damaged and can no longer function correctly, the entire body is affected with the following possible consequences [5, 6]:

- Uraemia or toxaemia caused by abnormal levels of waste products in the body
- Oedema, i.e. accumulation of water in tissues due to fluid overload
- Hypertension due to increased blood volume or a dysfunction of hormonal regulation
- Electrolytes imbalance leading to disorders in the cardiovascular or nervous system
- Hormonal changes (e.g. impaired production of erythropoietin leading to anaemia)

Table 1.1 Estimated number of worldwide patients with chronic kidney failure in 2017 [7]

	2017	Share (%)
Patients with chronic kidney failure	3,920,000	100
Patients with transplants	760,000	19
Dialysis patients	3,160,000	81
Haemodialysis (HD)	2,810,000	72
Peritoneal dialysis (PD)	350,000	9

- Abnormal enzyme production (e.g. high renin production)
- Decalcification of bones due to impaired production of calcitriol

CKD increases also the risk of cardiovascular disease and most of patients with CKD die from cardiovascular disease rather than kidney failure [9, 10].

Patients with severely damaged kidneys have to be treated through renal replacement therapy (RRT). The best treatment for patients with ESRD is a kidney transplant from a living or deceased donor, which allows the patient to lead a relatively normal life. However, given that the number of organs donated worldwide is significantly lower than the number of patients on transplant waiting lists, less than 20% of patients with chronic kidney failure receive kidney transplantation, and hence more than 80% of patients require dialysis treatment to substitute the function of the kidneys (see Table 1.1) [7].

Dialysis (from Greek διάλυσις, dialysis, meaning dissolution or separation) consists in therapeutic purification of blood and hence replaces the most important, life-saving functions of a healthy kidney, i.e.:

- Removing waste products of metabolism and toxic chemicals
- Removing excess water and salts
- Maintaining the correct levels of electrolytes in the body
- Correcting the metabolic acidosis
- Maintaining the correct level of arterial BP

Dialysis, therefore, does not replace all kidney functions (in particular, it does not substitute endocrine or metabolic activities of the kidneys), and hence it is usually supported by synthetic or recombinant hormones (such as erythropoietin), vitamins (such as vitamin D) or other medicines (e.g. phosphate binders) [8, 11]. Dialysis patients usually require also some dietary restrictions related to protein, electrolyte and fluid intake in order to reduce the interdialytic weight gain to minimum [8, 11].

The two common dialysis treatments are:

1. Haemodialysis (HD) – an extracorporeal technique in which the patient's blood flows to an external machine, where it is cleared from unwanted solutes and excess water in a dialyzer and then is returned to the body
2. Peritoneal dialysis (PD) – an intracorporeal technique in which a sterile solution containing an osmotic agent (e.g. glucose) is infused into the abdominal cavity, where it absorbs excess body water, waste products of metabolism and other solutes through a complex transport system formed by perfused capillaries immersed in the peritoneal tissue and is then drained out of the body via a catheter

HD is the most common dialysis modality with almost 90% of dialysis patients treated with this therapy. Dialysis patients are treated on a regular basis (maintenance dialysis) in over 40,000 dialysis centres worldwide [7]. Some patients, especially patients on PD, are also treated at home (home dialysis) [8, 11].

The number of dialysis patients worldwide is rising at a relatively constant rate of around 6% annually (a considerably higher rate than the world population growth) and was expected to reach around 3.4 million patients in 2018 [7] (Table 1.2). In highly developed regions, such as the USA, Western and Central Europe or Japan,

Table 1.2 Expected growth in dialysis patient numbers in 2018 [7]

Region	Expected growth in 2018 (%)
North America	~4
Europe, the Middle East and Africa	~4
Asia-Pacific	~9
Latin America	~4
Worldwide	~6

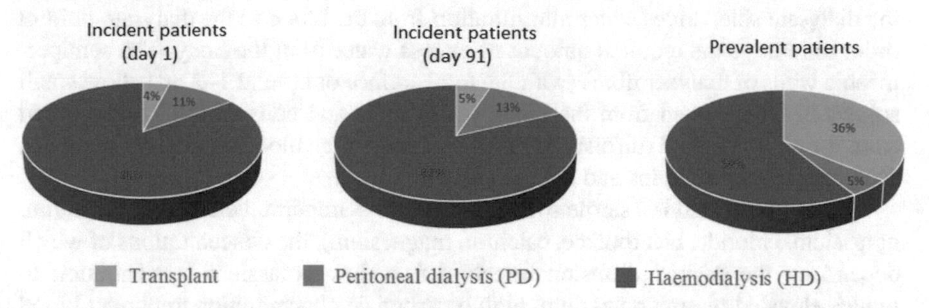

Fig. 1.2 Percentages of RRT modality in Europe in 2015 for incident patients (at the start of therapy and by day 91 of commencing RRT) and prevalent patients [12]

where the number of patients with ESRD is already relatively high and treatment is readily available to patients, the annual growth rate is below 4%, whereas in other, less-developed parts of the world, the growth rates are accordingly higher [7]. In the long term, given that the kidney function deteriorates with age, whereas the life expectancy continues to rise worldwide, as well as due to the increasing prevalence of lifestyle diseases such as diabetes or hypertension, which can impair kidney function and often precede the onset of ESRD, combined with the gradually improving access to medical care around the globe and the decreasing mortality among dialysis patients, the number of patients undergoing dialysis therapy is expected to increase from 3.2 million in 2017 to 4.9 million in 2025 (a rise of over 50%) [7]. It is estimated that the global dialysis market is currently worth around €70 billion, which is expected to increase by around 4% per year [7].

In Europe in 2015 over 81,000 patients commenced RRT for ESRD (an overall unadjusted incidence rate of 119 per million population) [12]. At the end of 2015, there were almost 550,000 patients on RRT (unadjusted prevalence rate of 801 pmp), of which 60% were men [12]. Most of these patients were on regular dialysis (59% on HD, 5% on PD), whereas 36% were living with a kidney transplant (see Fig. 1.2) [12].

1.3 Haemodialysis

In HD some of the patient's blood is diverted from the body (typically at a rate of 200–400 mL/min) to an extracorporeal circuit with a dialyzer, where excess water, waste products and toxins are removed from the blood which is then returned to the patient's bloodstream.

A dialyzer, or an artificial kidney, serves as a semipermeable membrane between the blood and dialysate fluid, acting as a blood filter, thus substituting the function of renal glomeruli and tubules. The contemporary dialyzers are hollow-fibre dialyzers with the blood flowing through a cylindrical bundle of thousands of porous fibres and the dialysate fluid washing the fibres externally in a countercurrent flow in order to keep the solute concentration gradient across the membrane at maximum for the best efficiency of dialysis. A hydraulic pressure difference between the blood and dialysate fluid, controlled by the dialysis machine by applying negative pressure on the dialysate side, drives water ultrafiltration from the blood to the dialysate fluid in order to remove the required amount of excess water from the body. The semipermeable walls of dialyzer fibres (with the total surface of around 1–2 m^2) allow small solutes to be extracted from the blood (diffusively and convectively) or absorbed from the dialysate fluid (diffusively) while leaving in the blood its vital components, such as proteins, vitamins and blood cells.

The dialysate fluid is a sterile solution of the most important electrolytes (sodium, potassium, chloride, bicarbonate, calcium, magnesium), the concentrations of which depend on the desired diffusion direction (e.g. low potassium concentration to reduce elevated plasma potassium, high bicarbonate concentration to correct blood acidity, normal sodium concentration to prevent excessive sodium loss from plasma, etc.).

Most patients on HD receive intermittent in-centre treatment with typically three dialysis sessions per week lasting 3–5 hours each with the ultrafiltration in the range of a few litres. For patients receiving home HD, the treatment frequency and time can be adjusted to meet individual patient needs (e.g. short daily sessions lasting only a couple of hours).

A haemodialysis system is a complex set of equipment that includes the bloodlines, blood pumps, dialyzer, dialysate solution or concentrate, water purification system (to produce dialysate fluid from the concentrate), control system (to regulate flow rates, dialysate conductivity, temperature, pH etc.), data processing system, safety monitoring system (BP monitor, clot and air bubble trap) and ports for administering drugs. During the treatment, an anticoagulant (most commonly heparin) has to be administered to avoid blood clotting in the dialyzer circuit.

The dialysis prescription includes the frequency and length of the treatment, the flow rates of blood and dialysate fluid, the type of dialyzer and the composition of the dialysate fluid.

Before each dialysis session, the dialyzer and the whole extracorporeal circuit are rinsed and primed with a sterile saline. At the beginning of the session, once the patient is connected to the extracorporeal system, the circuit is filled with the patient's blood (at a relatively slow flow rate of 50–100 mL/min), thus forcing out the priming saline, which can be either infused to the patient's bloodstream or discarded into a special drain bag [8].

The blood is taken and returned to the body via a special vascular access providing an adequate blood flow, which in most cases is an arteriovenous fistula – a surgical connection (anastomosis) of a vein and an artery (typically in the arm or forearm, e.g. between the cephalic vein and the radial artery) creating a vascular

Table 1.3 Types of vascular access for haemodialysis [8, 11]

	Principle	Advantages	Drawbacks
AV fistula	A vein and an artery are surgically joined together	High blood flow, low risk of infection, usually long-lasting	Takes several weeks to mature
AV graft	A small tube (typically synthetic) is inserted between a vein and an artery	High blood flow, usable in vessels not suitable for a fistula	Susceptible to clogging or infection
Central venous catheter	A double-lumen tube is inserted into one of the large veins in the chest or in the neck	Quick access to high volume bloodstream	A much greater risk of infection or clogging

shunt with a relatively fast and high blood flow. An AV fistula requires several weeks to fully develop before it can be used for treatment, but it is usually the first and best choice for dialysis access due to a relatively low risk of infection and its long lifetime (provided it is well maintained and monitored). Other types of vascular access for dialysis include an AV graft or a central venous catheter (see Table 1.3).

The removal of waste products and toxins during HD takes place predominantly by diffusion. The easiest to dialyse are low-molecular-weight, water-soluble molecules with low protein-binding capacity, such as urea and creatinine.

Depending on the size of the membrane pores, dialyzers are divided into low-flux or high-flux dialyzers. The latter are characterized by highly permeable membranes, which on one hand facilitate clearance of mid-molecular-weight molecules (such as β_2-microglobulin) but on the other hand enable the passage of some albumin molecules, which is generally undesirable.

As an alternative to conventional HD, other dialysis modalities may be used, such as haemodiafiltration (HDF) or haemofiltration (HF). In HDF, blood is also pumped through a dialyzer, but a higher pressure gradient is applied across the dialyzer membrane, which results in a considerably higher ultrafiltration rate than in HD and a more efficient removal of larger toxins by convection (the large amounts of water and salts removed from the blood in this process are replaced by infusing a substitution fluid). In HF, no dialysate fluid is present in the dialyzer, and hence the toxins are removed predominantly by convection.

1.4 Clinical Problems in Haemodialysis

Haemodialysis, or dialysis in general, is a life-saving and life-prolonging treatment for patients with ESRD; however, it does not fully replace the kidneys, nor does it reverse kidney failure or the underlying diseases. Therefore, despite dialysis treatment, ESRD patients are still subject to serious consequences of CKD (see Sect. 1.2), in particular to cardiovascular complications, which remain the leading cause of death in dialysed patients [9, 10].

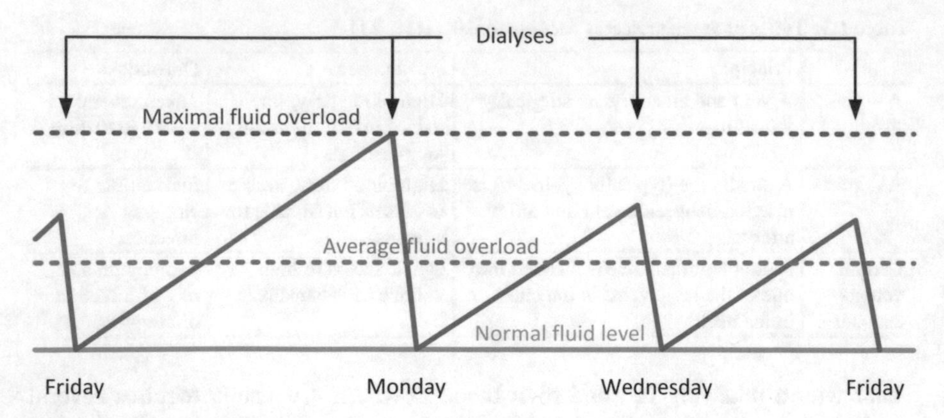

Fig. 1.3 Typical profile of fluid overload in patients on thrice-weekly maintenance haemodialysis

One of the main goals of HD is to restore the normal (or close to normal) fluid status of the patient and hence to remove the excess fluid accumulated in the body since the previous dialysis session and to reduce the patient's weight to the so called dry weight (the normal body weight). However, soon after the dialysis stops, the patient starts to accumulate fluid again, and hence, apart from the short period of time at the end and immediately after the dialysis session, patients on dialysis are in a constant fluid overload state (see Fig. 1.3), which is one of the main causes of cardiovascular disorders seen in such patients. Note that a dialysis with a preventively excessive ultrafiltration (below the patient's dry weight) is not a feasible option, as this would normally activate the thirst mechanism and the patient would instinctively drink a lot fluids immediately after dialysis, thus quickly restoring the lost fluid volume (potentially even above the dry weight level).

Haemodialysis procedure itself imposes a significant burden on the cardiovascular system and can generate a number of clinical problems during or after dialysis session, some of which can also impact the patient's in the long term, as discussed later in the chapter.

During a standard HD session, a few litres of fluid are removed from the body over a relatively short period of time (typically 3–5 hours). As this volume is removed in the dialyzer directly from the blood, thus reducing the circulatory volume, the functioning of the cardiovascular system relies on efficient vascular refilling [13–15] and the activity of the autonomic nervous system regulating the vascular tone and cardiac action. The majority of dialysed patients can comfortably deal with the reduction in blood volume and have their BP modestly reduced or unchanged [16, 17]. For some patients, however, the relatively rapid changes in the fluid status during HD can cause several problems, as discussed below.

1.4.1 *Intradialytic Hypotension*

Approximately 30% of dialysed patients have an insufficient cardiovascular response to the reduction in plasma volume and osmolarity during HD and cannot efficiently maintain BP, thus experiencing hypotension. Intradialytic hypotension (IDH) is the most frequent and problematic complication of HD [18]. The US National Kidney Foundation's Kidney Disease Outcomes Quality Initiative (KDOQI) defined IDH as a decrease of systolic blood pressure (SBP) of at least 20 mmHg or a decrease of mean arterial pressure (MAP) of at least 10 mmHg associated with symptoms [19]. In most studies, IDH is defined as a reduction of SBP below a certain threshold value or by a certain amount with or without associated symptoms (see Table 1.4). Some authors define IDH based not only on the magnitude of BP drop but also on the number of IDH episodes over a certain period of time [20].

IDH is usually characterized by the typical symptoms of hypotension, which may include dizziness, nausea, headache, abdominal discomfort, muscular cramps, vomiting, chills, fever, sweating, thirst, restlessness, cardiac palpitations, dyspnoea, anxiety or convulsions [19, 21]. A sudden and high decrease in arterial pressure can even lead to a circulatory collapse and a syncope [19, 22]. On the other hand, IDH may also occur asymptomatically [23].

Symptomatic IDH is not only highly uncomfortable for patients but is also problematic for the medical staff, as it significantly hinders the delivery of prescribed dialysis treatment, which in case of a severe hypotensive episode has to be either prematurely terminated or at least modified to limit further blood pressure decrease. The possible interventions include infusing saline or plasma expanders, reducing the dialyzer blood flow rate, reducing or stopping the ultrafiltration or performing the Trendelenburg or other manoeuvres [24]. All the above scenarios translate into a less effective dialysis treatment or under-dialysis [25].

In the long term, repetitive IDH can lead to cardiac or mesenteric ischaemia [26], cardiovascular instability [24], myocardial stunning [27, 28] or vascular access thrombosis [29]. Patients experiencing IDH have higher mortality rates [8], although there are contradictory data on whether IDH has an independent effect on mortality when other covariates are accounted for [30, 31].

Patients on HD can be divided into hypotension-prone and hypotension-resistant, but the susceptibility to IDH can vary from day to day in the same person. The

Table 1.4 Definitions of intradialytic hypotension seen in the literature (each row represents a separate definition based on systolic blood pressure (SBP) and other criteria)

SBP	SBP reduction	Other criteria	References
<90 mmHg	–	Hypotension symptoms	[34]
–	≥30 mmHg	Symptoms requiring an intervention	[21]
–	≥20 mmHg	Associated symptoms	[19]
<100 mmHg	–	–	[35]
<110 mmHg	>30 mmHg	–	[36]

Table 1.5 Incidence of intradialytic hypotension in different groups of patients

Frequency of IDH (%)	Comments	References
20	–	[37]
8–33	Age < 50: 8%; age 50–70: 19%; age > 70: 33%	[18]
10–30	–	[34, 38]
15–30	–	[39]
15–50	–	[40]
25–50	–	[41]
50	Patients with diabetes mellitus, autonomic insufficiency or cardiac disease; elderly patients	[38]
17–57	Critically ill patients, patients in ICU units	[42]

frequency of IDH depends on the studied group of patients (see Table 1.5) as well as on the rate of ultrafiltration [32] (it can be as low as 7% at the rate of 0.3 mL/min/kg or as high as 67% at the rate of 0.6 mL/min/kg [33]).

IDH can be caused by an excessive or too rapid decrease in blood volume (when ultrafiltration is too high compared to vascular refilling from the interstitial space), by impaired autonomic regulation (inadequate vasoconstriction or poor control of capacitance vessels), or other factors, such as diastolic dysfunction [8, 25]. The pathophysiology of IDH is multifactorial and not fully understood yet [43–45].

Several techniques are used to prevent IDH, such as performing more frequent or longer dialysis, lowering dialysate temperature and using variable concentration of sodium in the dialysate (sodium profiling) or variable ultrafiltration rate (ultrafiltration profiling) [39, 43]; however, since the mechanisms of IDH are complex and patient-specific, it remains a major problem in dialysis units [43].

1.4.2 *Intradialytic Hypertension*

Intradialytic hypertension – a paradoxical increase of BP during or immediately after HD – affects up to 15% of patients [46] and hence is generally less common than IDH. With no standard definition, different investigators define it as an increase of BP by a certain amount compared to the pre-dialysis value, e.g. an increase in MAP $\geq$ 15 mmHg [47] or an increase in SBP > 10 mmHg [48]. Similarly as for IDH, some authors define intradialytic hypertension on the basis of frequency of its occurrence over a certain period of time (e.g. over 6 months) [49, 50].

Intradialytic hypertension is associated with adverse outcomes, higher odds for hospitalization and higher mortality rates [49–51]. Factors increasing the risk of intradialytic hypertension include high fluid overload, high extracellular fluid overload (i.e. elevated ratio of extracellular to total body water), older age, cardiovascular comorbidities or short dialysis with low ultrafiltration [50].

The pathogenesis of intradialytic hypertension is multifactorial and poorly understood. Its likely causes include sympathetic overactivity (increased vasoconstriction), activation of the renin-angiotensin-aldosterone system (RAAS) or endothelial cell dysfunction [46, 50]. In order to prevent or minimise intradialytic hypertension, it is advisable to avoid the use of antihypertensive drugs that are easily dialysed, to use RAAS inhibitors and to reach adequate sodium removal during dialysis, which reduces interdialytic thirst and weight gain [46].

1.4.3 Disequilibrium Syndrome

The dialysis disequilibrium syndrome is a much less common but serious complication of HD, which still remains incompletely understood [52]. It involves a set of neurologic symptoms that may start from nausea, restlessness or headache but can lead to seizures, coma or even death [8, 52]. The most likely explanation of the disequilibrium syndrome is cerebral oedema due to dialysis-induced decrease in plasma osmolarity leading to the shift of water from plasma to brain tissue and cells [8, 52]. It is more likely to occur in highly uraemic patients undergoing their first dialysis treatment, if they are dialysed too quickly (a longer, slower and smaller dose of dialysis should be used in such cases) [8, 52]. For patients on regular maintenance dialysis, a possible approach to minimise the risk of the disequilibrium syndrome is to use dialysate sodium profiling, with a higher sodium concentration at the beginning of the session to counteract the reduction of plasma osmolarity due to intense urea removal [8]. Such an approach may, however, increase the interdialytic weight gain and BP [8].

1.4.4 Technical Issues

Haemodialysis is also associated with a number of possible technical complications during the dialysis session, such as clotting of the extracorporeal circuit, air embolism, access complications or blood leaks in case of blood lines disconnection or dialysis needles dislodging [53]. Even though the above should not occur in a correctly connected dialysis circuit with appropriate safety systems (such as the aforementioned air bubble trap), both the dialysed patient and the dialysis equipment should be carefully monitored by the medical staff during dialysis sessions.

1.5 Mathematical Models of Haemodialysis and Cardiovascular System

The dynamic exchange of water and solutes between different body fluid compartments, across the cellular membranes and capillary walls as well as through the flow of blood and lymph, involves a high number of variables and complex transport mechanisms, the interactions of which, even under physiological conditions, are difficult to predict or analyse without the aid of a mathematical model. Mathematical or computational modelling becomes even more important for analysing the relationships and interplay between different physiological mechanisms when the body systems are externally or internally perturbed and the whole body is out of its normal balance.

Haemodialysis is an excellent example of such external systemic perturbation. Even though it is aimed at restoring the natural body homeostasis (at least at the level of water-electrolyte balance, acidity and waste removal), it actually considerably disturbs the patient's body from its pre-dialysis state. Although the general idea of HD is relatively simple (to temporarily pass some of the blood through the extracorporeal circuit containing a semipermeable membrane), the processes taking place in the dialyzer, i.e. ultrafiltration and bidirectional solute transport, by changing the volume and composition of the circulating blood, affect literally all body fluid compartments and the whole cardiovascular system, from the volume of red blood cells or the hydrostatic pressure of the interstitial fluid to the peripheral vascular resistance and cardiac filling. The use of model-based computer simulations can enable a quantitative analysis of dialysis-induced processes taking place in the patient's body under different dialysis treatment scenarios in an attempt to better understand the physiological or pathophysiological mechanisms behind the observed cardiovascular responses to HD.

There exist a number of mathematical models describing water and solutes distribution and transport in the body, developed either to analyse blood volume changes during fluid resuscitation or fluid infusions [54–56] or to evaluate solute and water kinetics during haemodialysis [57–60]. Such models are typically multi-compartment models in which the analysed body fluid spaces of similar or same properties are lumped together into several compartments, e.g. all tissue cells lumped into one intracellular compartment or all red blood cells lumped into one RBC compartment[1] (see Fig. 1.4). Each of such compartments is described by a separate

[1] Distinguishing between plasma and red blood cell fluid (i.e. having a separate compartment for red blood cells) is related to the fact that both fluids differ in ion and protein concentrations and the fact that the red blood cell membrane provides a barrier for the passive transport of solutes between the two fluid spaces with different permeabilities to different solutes, which affects the transport processes during dialysis (when some solutes are removed or added to plasma), given that the red blood cells occupy almost half of the vascular space. Other blood cells, such as white blood cells or platelets, are typically ignored due to their relatively low total volume (<1%).

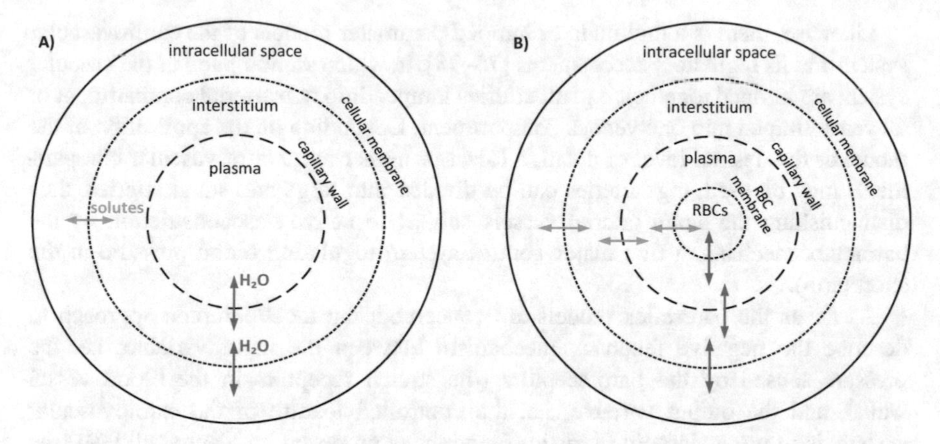

Fig. 1.4 Examples of multi-compartment models of whole-body water and solute transport. (**a**) A combination of a three-compartment model for the description of osmotic water shifts and a two-compartment model for the transport of low-molecular-weight solutes (in the latter, the interstitial fluid and blood plasma are treated as one extracellular compartment, assuming both have similar solute concentrations); (**b**) a four-compartment model with an additional red blood cell compartment, with water and solute exchange between all compartments (across three different barriers)[1]

set of parameters and the transport of matter between individual compartments is described using differential equations.

Modelling of fluid and solute distribution and transport in the body during and between dialysis sessions has a long history [8, 61–63]. In 1985 the analysis of clinical data using the simplest, one-compartment model of urea distribution and kinetics led to the definition of the first haemodialysis adequacy index KT/V, where KT (dialyzer clearance times duration of dialysis session) is the dialysis 'dose' and V is the urea distribution volume (i.e. total body water) reflecting the size of the patient [64]. The subsequent kinetic studies of haemodialysis with more efficient dialyzers (resulting in higher solute removal over a shorter time, thus causing a stronger perturbation of the body transport system) resulted in the introduction of a two-compartment model for urea and creatinine – the two most often used markers of dialysis adequacy and the development of semiempirical formulae for efficient estimation of KT/V from clinical data [65–67]. Several kinetic models for other solutes, such as vitamin B12, inulin, phosphate, sodium, potassium, bicarbonate, etc., were then developed. New adequacy indices were formulated following the appearance of new dialysis schedules, such as frequent dialysis, and the indices system was eventually applied for novel markers of dialysis adequacy, such as phosphate [68–72]. More recently, an alternative approach to modelling solute kinetics – a regional blood flow model – was formulated on the basis of the discrimination between highly perfused organs (such as heart, brain, kidneys or liver) and poorly perfused organs and tissues (such as muscles, skin or bones) [73, 74].

Likewise, there is a multitude of lumped-parameter models of the cardiovascular system and its regulatory mechanisms [75–78], in which similar parts of the vascular system are lumped together, e.g. all arteries lumped into one arterial compartment or all veins lumped into one venous compartment. Depending on the application of the model or the desired level of detail, a lower or higher number of vascular compartments may be used, e.g. arteries can be divided into large and small arteries, thus distinguishing the small arterial vessels subject to active vasoconstriction by the baroreflex mechanism (the major control system regulating blood pressure in the short term).

As far as the baroreflex models are concerned, the most common approach to describe the negative feedback mechanism between the input variable, i.e. the pressure sensed by the baroreceptors (the stretch receptors in the blood vessel walls), and the output variable, i.e. the controlled feature of the cardiovascular system, is to use a sigmoid (logistic) function, as proposed by Kent et al. [79] (see Fig. 1.5) with certain dynamics.

In a healthy human, the baroreflex controls the heart rate, heart contractility, peripheral vascular resistance and venous capacity (venous unstressed volume and venous compliance). The baroreflex regulation is effectuated through modulation of the activity of sympathetic and parasympathetic limbs of the autonomic nervous system based on blood pressure deviations from the reference level sensed by the arterial baroreceptors located in the aortic arch and carotid sinuses, or by the cardiopulmonary baroreceptors located in the atria, ventricles and pulmonary vessels. Depending on the purpose of the model, different authors use a more or less detailed description of the baroreflex mechanisms. Some authors take into account only arterial baroreceptors [80], whereas others include also the cardiopulmonary baroreceptors [81, 82]. Some models describe only the regulation of heart rate,

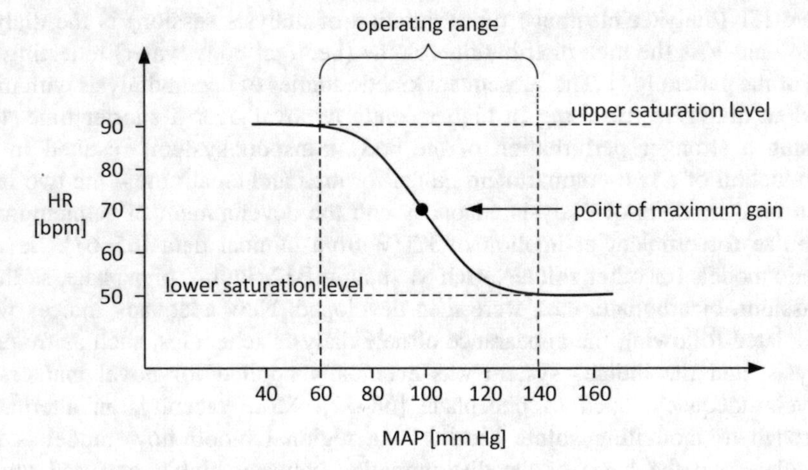

Fig. 1.5 The standard sigmoidal curve describing the static characteristics of the negative feedback baroreflex mechanism regulating the heart rate (output) based on mean arterial pressure (input). The gain or sensitivity of the reflex is maximal at the midpoint between the upper and lower saturation levels. The operating range is the range of MAP over which changes in MAP evoke changes in HR

peripheral resistance and venous unstressed volume, whereas other models include additionally the control of heart contractility [81, 83, 84] or venous compliance [75, 80].

Finally, several mathematical models were developed to describe the transport processes in the dialyzer during HD [85, 86].

Despite the existence of such a multitude of models of the cardiovascular system, baroreflex mechanisms, whole-body water and solute transport or the mass exchange in the dialyzer circuit, to the best knowledge of the authors, only one attempt has been made to integrate the above models in a single framework to study the arterial blood pressure during HD [82]. The model by Ursino and Innocenti [82] (the six-compartment model of the cardiovascular system with three baroreflex mechanisms controlling heart rate, systemic resistance and venous unstressed volume) was used for the analysis of the cardiovascular response to haemodialysis by reproducing a few patterns of mean arterial pressure changes during HD taken from the literature (including moderate and severe intradialytic hypotension). However, similarly to most of the aforementioned models, the model by Ursino and Innocenti [82] uses several important simplifications, such as treating all blood plasma as one vascular compartment exchanging fluid with the interstitium (thus neglecting the differences in plasma composition across the cardiovascular system, in particular the differences between plasma in the systemic capillaries and in the dialyzer, where intense water and solute exchange processes take place), assuming constant vascular resistances regardless of blood volume level and haematocrit level (apart from the resistance of systemic arteries being controlled by the baroreflex), ignoring the variation of haematocrit across the cardiovascular system or employing an out-of-date model of microvascular exchange of water and solutes (with arterial filtration balanced by venous reabsorption at the steady state). It also includes a very limited number of osmotic agents (sodium, potassium and urea) and ignores the lymphatic system or the transcapillary protein and small solute transport processes. Due to the much simplified approach to modelling water and solute transport within the body and a relatively simple structure of the modelled cardiovascular system, the model by Ursino and Innocenti [82], although useful for an approximate qualitative analysis of possible arterial blood pressure patterns during HD, cannot be used for a detailed analysis of pressure and volume changes of vascular and extravascular fluid compartments or changes in concentration of the most important osmotically active solutes (ions, low-molecular-weight solutes and proteins) across the human body both under physiological conditions and during HD. The above limitations are addressed in the new comprehensive model presented and discussed in this book.

References

1. Bhave, G., Neilson, E.G.: Body fluid dynamics: back to the future. J Am Soc Nephrol. **22**, 2166–2181 (2011)
2. Guyton, A.C., Hall, J.E.: Textbook of Medical Physiology, 11th edn. Elsevier Saunders, Philadelphia (2006)

3. Klahr, S. (ed.): The Kidney and Body Fluids in Health and Disease, 1st edn. Springer Science +Business Media, LLC., New York (1983)
4. Kidney Disease: Improving Global Outcomes (KDIGO) CKD Work Group: KDIGO 2012 clinical practice guideline for the evaluation and management of chronic kidney disease. Kidney Int Suppl. **3**(1), 1–150 (2013)
5. Drawz, P., Rahman, M.: Chronic kidney disease. Ann Intern Med. **162**(11), ITC1–IT16 (2015)
6. Hoffsten, P., Klahr, S.: Pathophysiology of chronic renal failure. In: Klahr, S. (ed.) The Kidney and Body Fluids. Springer Science + Business Media LLC, New York (1983)
7. Fresenius Medical Care: Annual Report 2017. Fresenius Medical Care AG & Co. KGaA, Bad Homburg (2018)
8. Daugirdas, J.T., Blake, G., Ing, T.S. (eds.): Handbook of Dialysis, 5th edn. Wolters Kluwer Health, Philadelphia (2015)
9. Santoro, A., Mandreoli, M.: Chronic renal disease and risk of cardiovascular morbidity-mortality. Kidney Blood Press Res. **39**(142), 142–146 (2014)
10. Di Lullo, L., House, A., Gorini, A., Santoboni, A., Russo, D., Ronco, C.: Chronic kidney disease and cardiovascular complications. Heart Fail Rev. **20**(3), 259–272 (2015)
11. Ahmad, S.: Manual of Clinical Dialysis, 2nd edn. Springer Science+Business Media, LLC., New York (2009)
12. ERA-EDTA Registry: ERA-EDTA Registry Annual Report 2015. ERA-EDTA Registry, Amsterdam (2017)
13. Schneditz, D., Roob, J., Oswald, M., Pogglitsch, H., Moser, M., Kenner, T., et al.: Nature and rate of vascular refilling during hemodialysis and ultrafiltration. Kidney Int. **42**(6), 1425–1433 (1992)
14. Pietribiasi, M., Katzarski, K., Galach, M., Stachowska-Pietka, J., Schneditz, D., Lindholm, B., et al.: Kinetics of plasma refilling during hemodialysis sessions with different initial fluid status. ASAIO J. **61**(3), 350–356 (2015)
15. Pietribiasi, M., Wojcik-Zaluska, A., Zaluska, W., Waniewski, J.: Does the plasma refilling coefficient change during hemodialysis sessions? Int J Artif Organs. **41**(11), 706–713 (2018b)
16. Tonelli, M., Astephen, P., Andreou, P., Beed, S., Lundrigan, P., Jindal, K.: Blood volume monitoring in intermittent hemodialysis for acute renal failure. Kidney Int. **62**, 1075–1080 (2002)
17. Chou, K.J., Lee, P.T., Chen, C.L., Chiou, C.W., Hsu, C.Y., Chung, H.M., et al.: Physiological changes during hemodialysis in patients with intradialysis hypertension. Kidney Int. **69**(10), 1833–1838 (2006)
18. Zucchelli, P., Santoro, A.: Dialysis-induced hypotension – a fresh look at pathophysiology. Blood Purif. **11**, 85–98 (1993)
19. K/DOQI Workgroup: K/DOQI clinical practice guidelines for cardiovascular disease in dialysis patients. Am J Kidney Dis. **45**(4 Suppl 3), S1–S153 (2005)
20. Atabak, S.: Change in definition could change results. Nephrol Dial Transplant. **19**(10), 2676 (2004)
21. Andrulli, A., Colzani, S., Mascia, F., Lucchi, L., Stipo, L., Bigi, M.C., et al.: The role of blood volume reduction in the genesis of intradialytic hypotension. Am J Kidney Dis. **40**(6), 1244–1254 (2002)
22. Palmer, B.F., Heinrich, W.L.: Recent advances in the prevention and management of intradialytic hypotension. J Am Soc Nephrol. **19**(1), 8–11 (2008)
23. Bradshaw, W., Bennett, P.N.: Asymptomatic intradialytic hypotension: the need for pre-emptive intervention. Nephrol Nurs J. **42**(5), 479–485 (2015)
24. Cavalcanti, S., Ciandrini, A., Severi, S., Badiali, F., Bini, S., Gattiani, A., et al.: Model-based study of the effects of the hemodialysis technique on the compensatory response to hypovolemia. Kidney Int. **65**(4), 1499–1510 (2004)
25. Daugirdas, J.T.: Pathophysiology of dialysis hypotension: an update. Am J Kidney Dis. **38** (4 Suppl 4), S11–S17 (2001)
26. Schreiber Jr., M.J.: Setting the stage. Am J Kidney Dis. **38**(4 Suppl 4), S1–S10 (2001)

27. McIntyre, C.W.: Haemodialysis-induced myocardial stunning in chronic kidney disease – a new aspect of cardiovascular disease. Blood Purif. **29**(2), 105–110 (2010a)
28. McIntyre, C.W.: Recurrent circulatory stress: the dark side of dialysis. Semin Dial. **23**(5), 449–451 (2010b)
29. Chang, T.I., Paik, J., Greene, T., Desai, M., Bech, F., Cheung, A.K., et al.: Intradialytic hypotension and vascular access thrombosis. J Am Soc Nephrol. **22**, 1526–1533 (2011)
30. Shoji, T., Tsubakihara, Y., Fujii, M., Imai, E.: Hemodialysis-associated hypotension as an independent risk factor for two-year mortalty in hemodialysis patients. Kidney Int. **66**, 1212–1220 (2004)
31. Tislér, A., Akócsi, K., Borbás, B., Fazakas, L., Ferenczi, S., Görögh, S., et al.: The effect of frequent or occasional dialysis-associated hypotension on survival of patients on maintenance haemodialysis. Nephrol Dial Transplant. **18**, 2601–2605 (2003)
32. Sherman, R.A.: Intradialytic hypotension – an overview of recent, unresolved and overlooked issues. Sem Dial. **15**(3), 141–143 (2002)
33. Ronco, C., Feriani, M., Chiaramonte, S., Conz, P., Brendolan, A., Bragantini, L., et al.: Impact of high blood flows on vascular stability in haemodialysis. Nephrol Dial Transplant. **5**(Suppl 1), 109–114 (1990)
34. Letteri, J.M.: Symptomatic hypotension during hemodialysis. Sem Dial. **11**(4), 253–256 (1998)
35. Oliver, M.J., Edwards, L.J., Churchill, D.N.: Impact of sodium and ultrafiltration profiling on hemodialysis-related symptoms. J Am Soc Nephrol. **12**(1), 151–156 (2001)
36. Imai, E., Fujii, M., Kohno, Y., Kageyama, H., Nakahara, K., Hori, M., et al.: Adenosine A1 receptor antagonist improves intradialytic hypotension. Kidney Int. **69**(5), 877–883 (2006)
37. Daugirdas, J.T.: Preventing and managing hypotension. Sem Dial. **7**(4), 276–283 (1994)
38. Perazella, M.A.: Approach to patients with intradialytic hypotension: a focus on therapeutic options. Sem Dial. **12**(3), 175–181 (1999)
39. Reilly, R.F.: Attending rounds: a patient with intradialytic hypotension. Clin J Am Soc Nephrol. **9**, 798–803 (2014)
40. Orofino, L., Quereda, C., Villafruela, J.J., Sabater, J., Matesanz, R., et al.: Epidemiology of symptomatic hypotension in hemodialysis – is cool dialysate beneficial for all patients? Am J Nephrol. **10**(3), 177–180 (1990)
41. Henrich, W.L.: Hemodynamic instability during hemodialysis. Kidney Int. **30**(4), 605–612 (1986)
42. Bitker, L., Bayle, F., Yonis, H., Gobert, F., Leray, V., Taponnier, R., et al.: Prevalence and risk factors of hypotension associated with preload-dependence during intermittent hemodialysis in critically ill patients. Crit Care. **20**, 44 (2016)
43. Bradshaw, W.: Intradialytic hypotension: a literature review. Renal Soc Aust J. **10**(1), 22–29 (2014)
44. Reeves, P.B., Causland, M., F, R.: Mechanisms, clinical implications, and treatment of intradialytic hypotension. Clin J Am Soc Nephro. **13**, CJN.12141017 (2018)
45. Henrich, W.L.: Intradialytic hypotension: a new insight to an old problem. Am J Kidney Dis. **52**(2), 209–210 (2008)
46. Inrig, J.K.: Intradialytic hypertension: a less-recognized cardiovascular complication of hemodialysis. Am J Kidney Dis. **55**(3), 580–589 (2010)
47. Amerling, R.G., Dubrow, A., Levin, N.W., Osheroff, R. (eds.): Complications During Hemodialysis. Appleton and Lange, Stamford (1995)
48. Inrig, J.K., Patel, U.D., Toto, R., Szczech, L.A.: Association of blood pressure increases during hemodialysis with 2-year mortality in incident hemodialysis patients: a secondary analysis of the dialysis morbidity and mortality wave 2 study. Am J Kidney Dis. **54**(5), 881–890 (2009)
49. Van Buren, P.N., Inrig, J.K.: Mechanisms and treatment of intradialytic hypertension. Blood Purif. **41**(1–3), 188–193 (2016)
50. Van Buren, P.N., Inrig, J.K.: Special situations: intradialytic hypertension/chronic hypertension and intradialytic hypotension. Semin Dial. **30**(6), 545–552 (2017)
51. Choi, C.Y., Park, J.S., Yoon, K.T., Gil, H.W., Lee, E.Y., Hong, S.Y.: Intra-dialytic hypertension is associated with high mortality in hemodialysis patients. PLoS One. **12**(7), e0181060 (2017)

52. Zepeda-Orozco, D., Quigley, R.: Dialysis disequilibrium syndrome. Pediatr Nephrol. **27**(12), 2205–2211 (2012)
53. Smeltzer, S.C., Bare, B., Hinkle, J.L., Cheever, K.H.: Management of patients with renal disorders. In: Brunner & Suddarth's Textbook of Medical-Surgical Nursing, vol. 1, 12th edn. LWW, Philadelphia (2010)
54. Gyenge, C.C., Bowen, B.D., Reed, R.K., Bert, J.L.: Transport of fluid and solutes in the body I. Formulation of a mathematical model. Am J Physiol Heart Circ Physiol. **277**(3 Pt 2), H1215–H1227 (1999)
55. Wolf, M.B.: Whole body acid-base and fluid-electrolyte balance: a mathematical model. Am J Physiol Renal Physiol. **305**(8), F1118–F1131 (2013)
56. Bert, J.L., Gyenge, C.C., Bowen, B.D., Reed, R.K., Lund, T.: A model of fluid and solute exchange in the human: validation and implications. Acta Physiol Scand. **170**(3), 201–209 (2000)
57. Ursino, M., Coli, L., Brighenti, C., Chiari, L., de Pascalis, A., Avanzolini, G.: Prediction of solute kinetics, acid-base status, and blood volume changes during profiled hemodialysis. Ann Biomed Eng. **28**(2), 204–216 (2000)
58. Coli, L., Ursino, M., De Pascalis, A., Brighenti, C., Dalmastri, V., La Manna, G., et al.: Evaluation of intradialytic solute and fluid kinetics. Setting up a predictive mathematical model. Blood Purif. **18**(1), 37–49 (2000)
59. Pietribiasi, M., Waniewski, J., Załuska, A., Załuska, W., Lindholm, B.: Modelling transcapillary transport of fluid and proteins in hemodialysis patients. PLoS One. **11**(8), e0159748 (2016)
60. Pietribiasi, M., Waniewski, J., Wojcik-Zaluska, A., Zaluska, W., Lindholm, B.: Model of fluid and solute shifts during hemodialysis with active transport of sodium and potassium. PLoS ONE. **13**(12), e0209553 (2018)
61. Hörl, W.H., Koch, K.M., Lindsay, R.M., Ronco, C., Winchester, J.F. (eds.): Replacement of Renal Function by Dialysis, 5th edn. Springer-Science+Business Media, Dordrecht (2004)
62. Gotch, F.A., Sargent, J.A., Keen, M.L., Lee, M.: Individualized, quantified dialysis therapy of uremia. Proc Clin Dial Transplant Forum. **4**, 27–35 (1974)
63. Sargent, J.A.: Control of dialysis by a single-pool urea model: the National Cooperative Dialysis Study. Kidney Int Suppl. **13**, S19–S25 (1983)
64. Gotch, F.A., Sargent, J.A.: A mechanistic analysis of the National Cooperative Dialysis Study (NCDS). Kidney Int. **28**(3), 526–534 (1985)
65. Lowrie, E.G.: The kinetic behaviors of urea and other marker molecules during hemodialysis. Am J Kidney Dis. **50**(2), 181–183 (2007)
66. Daugirdas, J.T.: Second generation logarithmic estimates of single-pool variable volume Kt/V: an analysis of error. J Am Soc Nephrol. **4**(5), 1205–1213 (1993)
67. Ziolko, M., Pietrzyk, J.A., Grabska-Chrzastowska, J.: Accuracy of hemodialysis modeling. Kidney Int. **57**(3), 1152–1163 (2000)
68. Depner, T., Beck, G., Daugirdas, J., Kusek, J., Eknoyan, G.: Lessons from the Hemodialysis (HEMO) Study: an improved measure of the actual hemodialysis dose. Am J Kidney Dis. **33**(1), 142–149 (1999)
69. Leypoldt, J.K.: Kinetics of beta2-microglobulin and phosphate during hemodialysis: effects of treatment frequency and duration. Semin Dial. **18**(5), 401–408 (2005)
70. Leypoldt, J.K., Cheung, A.K., Deeter, R.B., Goldfarb-Rumyantzev, A., Greene, T., Depner, T. A., et al.: Kinetics of urea and beta-microglobulin during and after short hemodialysis treatments. Kidney Int. **66**(4), 1669–1676 (2004)
71. Waniewski, J., Debowska, M., Lindholm, B.: Theoretical and numerical analysis of different adequacy indices for hemodialysis and peritoneal dialysis. Blood Purif. **24**(4), 355–366 (2006)
72. Debowska, M., Waniewski, J., Lindholm, B.: An integrative description of dialysis adequacy indices for different treatment modalities and schedules of dialysis. Artif Organs. **31**(1), 61–69 (2007)
73. Schneditz, D., Van Stone, J.C., Daugirdas, J.T.: A regional blood circulation alternative to in-series two compartment urea kinetic modeling. ASAIO J. **39**(3), M573–M577 (1993)

74. Schneditz, D., Platzer, D., Daugirdas, J.T.: A diffusion-adjusted regional blood flow model to predict solute kinetics during haemodialysis. Nephrol Dial Transplant. **24**, 2218–2224 (2009)
75. Ottesen, J.T., Olufsen, M.S., Larsen, J.K. (eds.): Mathematical Models in Human Physiology. SIAM, Philadelphia (2004)
76. Olufsen, M.S., Ottesen, J.T., Tran, H.T., Ellwein, L.M., Lipsitz, L.A., Novak, V.: Blood pressure and blood flow variation during postural change from sitting to standing: model development and validation. J Appl Physiol. **99**(4), 1523–1537 (2005)
77. Heldt, T.: Computational models of cardiovascular response to orthostatic stress. PhD thesis, Massachusetts Institute of Technology (2004)
78. Pstras, L., Thomaseth, K., Waniewski, J., Balzani, I., Bellavere, F.: Mathematical modelling of cardiovascular response to the Valsalva manoeuvre. Math Med Biol. **34**(2), 261–292 (2017)
79. Kent, B.B., Drane, J.W., Blumenstein, B., Manning, J.: A mathematical model to assess changes in the baroreceptor reflex. Cardiology. **57**(5), 295–310 (1972)
80. Ursino, M., Antonucci, M., Belardinelli, E.: Role of active changes in venous capacity by the carotid baroreflex: analysis with a mathematical model. Am J Physiol Heart Circ Physiol. **267** (6), H2531–H2546 (1994)
81. Magosso, E., Biavati, V., Ursino, M.: Role of the baroreflex in cardiovascular instability: a modeling study. Cardiovasc Eng. **1**(2), 101–115 (2001)
82. Ursino, M., Innocenti, M.: Modeling arterial hypotension during hemodialysis. Art Org. **21**(8), 873–890 (1997)
83. Ursino, M.: Interaction between carotid barregulation and the pulsating heart: a mathematical model. Am J Physiol. **275**(5), H1733–H1747 (1998)
84. Ursino, M., Magosso, E.: Role of short-term cardiovascular regulation in heart period variability: a modeling study. Am J Physiol Heart Circ Physiol. **284**(4), H1479–H1493 (2003)
85. Sargent, J.A., Gotch, F.A.: Principles and biophysics of dialysis. In: Jacobs, C., Kjellstrand, C. M., Koch, K.M., Winchester, J.F. (eds.) Replacement of Renal Function by Dialysis, 4th edn, pp. 34–102. Springer Netherlands, Dordrecht (1996)
86. Waniewski, J.: Mathematical modeling of fluid and solute transport in hemodialysis and peritoneal dialysis. J Membrane Sci. **274**, 24–37 (2006)

Chapter 2
Integrated Model of Cardiovascular System, Body Fluids and Haemodialysis Treatment: Structure, Equations and Parameters

Abstract This chapter presents a mathematical model of the cardiovascular system with baroreflex regulation integrated with the model of the whole-body transport of water and solutes (ions, small molecules and proteins) during haemodialysis treatment. The chapter provides a detailed description of the proposed model structure and modelling approach with all the equations used, the assumptions made, the values of model parameters and the definition of initial steady-state conditions for both a healthy subject and a reference patient before dialysis.

Keywords Compartmental modelling · Model structure · Cardiovascular system · Vascular resistance · Microcirculation · Haematocrit · Fåhraeus effect · Cardiac output · Afterload · Heart rate · Baroreflex · Baroreceptors · Body fluids · Intracellular fluid · Extracellular fluid · Solute kinetics · Transcapillary fluid transport · Lymphatic system · Cellular membrane transport · Dialyzer · Dialysis access · Haemodialysis Modelling · Parameter assignment · Initial conditions

2.1 Overview

The presented model describes the two-phase blood flow (plasma with suspended red blood cells) across nine cardiovascular compartments and three extracorporeal compartments, as well as the water and solute transport between the vascular and extravascular compartments (across the capillary wall), between the extracellular and intracellular compartments (across the tissues cell membrane) and between plasma and red blood cells (across the red blood cell membrane). The model involves baroreflex regulatory mechanisms controlling heart rate, heart contractility, resistance of arterioles and venous unstressed volume, based on blood pressures recorded by the baroreceptors located in the right atrium (representing the low-pressure cardiopulmonary baroreceptors) and in large arteries (representing the high-pressure arterial baroreceptors). The model includes also the absorption of the interstitial fluid by the lymphatic system and its subsequent drainage to the systemic veins.

L. Pstras, J. Waniewski, *Mathematical Modelling of Haemodialysis*,
https://doi.org/10.1007/978-3-030-21410-4_2

The model can be subdivided into two interrelated layers: (I) the cardiovascular system (CVS) with its regulatory mechanisms and (II) transport of water and solutes within the body and in the dialyzer circuit.

The general structure of the cardiovascular part of the model is similar to the model developed earlier by the authors for simulating the Valsalva manoeuvre [1] (the VM model [2, 3]), with a few modifications discussed later in the chapter. The most important differences compared to other similar CVS models [4, 5] are as follows: (1) introduction of additional vascular compartments and extracorporeal circuit compartments, (2) use of a nonlinear (instead of linear) relationship between atrial pressure and stroke volume (Frank-Starling mechanism), (3) a new approach to modelling the impact of afterload on cardiac output, (4) introduction of the relationship between the vessel volume and its resistance to blood flow, (5) introduction of the dependence of blood viscosity on haematocrit (HCT) and (6) allowance for HCT variation across the CVS. The above features of the cardiovascular system were introduced to describe the well-known physiological phenomena and hence to enable a more detailed analysis of haemodynamic changes during HD.

Compared to previously formulated models of whole-body transport of water and solutes [6–12], the proposed model includes the following new features: (1) use of separate plasma and red blood cell (RBC) compartments for each cardiovascular compartment with dynamic water and solute exchange across the RBC membrane (instead of one lumped plasma compartment and one lumped RBC compartment for the whole body) – to improve the description of solute kinetics in blood, local and global changes of haematocrit as well as changes in vascular resistance; (2) extended description of small solutes transport across the capillary wall – to account for the impact of transcapillary osmotic pressure resulting from unequal distribution of ions and delayed transcapillary transport of some small solutes during dialysis; and (3) description of vascular leakage and refilling separately for albumin and non-albumin proteins – to better describe the flow of proteins in the modelled system and to enable analysis of the relative impact of those two groups of proteins on plasma and interstitial oncotic pressure changes during HD.

2.2 Cardiovascular System

2.2.1 Cardiovascular Compartments

The CVS is modelled by nine lumped compartments connected in series (see Fig. 2.1) corresponding to seven vascular compartments (large arteries, small arteries, capillaries, small veins, large veins, pulmonary arteries and pulmonary veins) and two cardiac compartments (right heart and left heart, each combining the corresponding atrium and ventricle). The model includes also an arteriovenous fistula (a vascular access for HD), described as a pure resistance (constant) placed between large arteries and large veins (see Fig. 2.1).

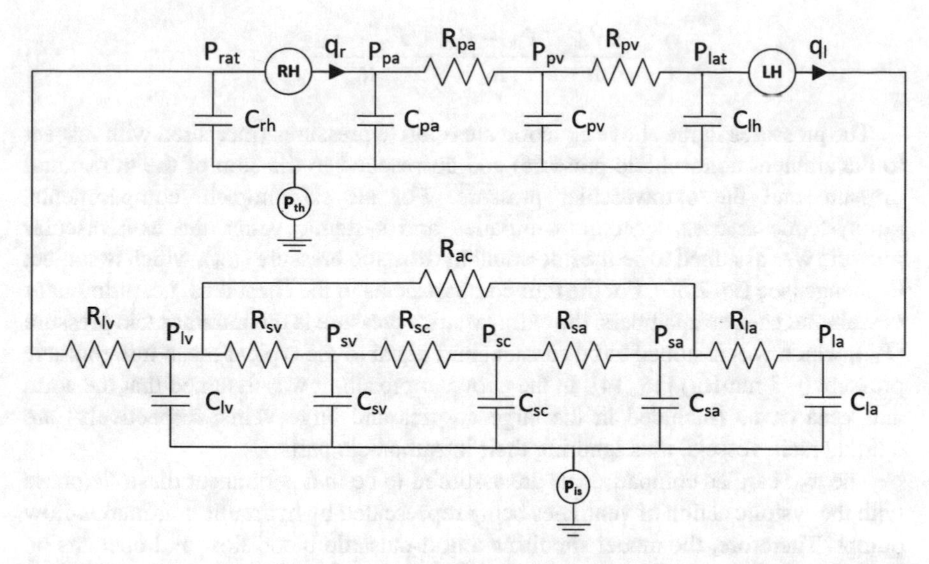

Fig. 2.1 Electric analogy of the cardiovascular model, where R denotes vascular resistances, P, pressures; C, vascular capacitances; q_r and q_l, cardiac output from right and left heart ventricles, respectively. Subscripts: la, large arteries; sa, small arteries; sc, systemic capillaries; sv, small veins; lv, large veins; pa, pulmonary arteries; pv, pulmonary veins; rh, right heart; lh, left heart; rat, right atrium; lat, left atrium; ac, arteriovenous access; th, intrathoracic; is, interstitial

Each vascular compartment is modelled as a capacitor (representing the volume of blood stored in the compartment at a given pressure) with hydraulic resistances between the compartments corresponding to pressure and energy losses associated with the blood flow.

Applying the law of mass conservation, the volume changes of the i-th compartment are calculated from the difference between the blood inflow ($Q_{\text{in},i}$) and outflow ($Q_{\text{out},i}$) of these compartments:

$$\frac{\mathrm{d}V_i}{\mathrm{d}t} = Q_{\text{in},i} - Q_{\text{out},i} \tag{2.1}$$

where the blood flows are given either by the output of cardiac chambers (see Sect. 2.2.3) or correspond to the flows driven by the pressure difference between the adjacent vascular compartments A and B (counteracted by hydraulic resistance of compartment A) determined analogously to Ohm's law for electrical current with the blood flow, pressure and hydraulic resistance being equivalent to current, voltage and electrical resistance, respectively:

$$Q_{\text{A-B}} = \frac{P_\text{A} - P_\text{B}}{R_\text{A}} \tag{2.2}$$

For instance, for the small arteries compartment the following equation applies:

$$\frac{dV_{sa}}{dt} = \frac{P_{la} - P_{sa}}{R_{la}} - \frac{P_{sa} - P_{sc}}{R_{sa}} \tag{2.3}$$

The pressures in the above equation are relative pressures (measured with respect to the ambient atmospheric pressure) and correspond to the sum of the transmural pressure and the extravascular pressure. For all extrathoracic compartments, i.e. systemic arteries, systemic capillaries and systemic veins, the extravascular pressure was assumed to be the interstitial hydrostatic pressure (P_{is}), which is subject to change (see Eq. 2.50). For the four compartments in the chest area, i.e. pulmonary vessels and cardiac chambers, the extravascular pressure is the intrathoracic pressure (P_{th}), which was assumed to be constant and equal to the typical mean intrathoracic pressure (−3 mmHg) [13, 14]. In the above approach, it was assumed that the aorta and vena cavae (included in the large arteries and large veins, respectively) are extrathoracic vessels, thus ignoring their intrathoracic parts.

The two cardiac compartments are assumed to be in a permanent diastolic phase with the systolic action of ventricles being represented by hydraulic continuous-flow pumps. Therefore, the model simulates a non-pulsatile blood flow and operates on mean blood pressures. The inertance effects were neglected due to minor relevance in a non-pulsatile blood flow model [15].

The VM model [2] had a separate aortic compartment and a separate vena cavae compartment. In the present model, the arterial tree has been divided into large arteries (internal diameter > 2.5 mm, including aorta) and small arteries (including arterioles). Similarly, the venous bed has been divided into large veins (internal diameter > 1 mm, including vena cavae) and small veins (including venules). There are three reasons for the above modifications. Firstly, having separate compartments for aorta and vena cavae was not needed in this study (unlike in modelling of the Valsalva manoeuvre [2, 16] where this was crucial). Secondly, with the division between large and small vessels, it is possible to account for the variation of HCT between different compartments (discussed later) and to introduce an arteriovenous fistula. Finally, such a division will enable in the future differentiating between blood flows to different tissues and organs (e.g. highly perfused vs lightly perfused organs [17]) with the subcirculations starting and ending in large arteries and large veins, respectively.

Given the considerably different pressures between the systemic arteries and capillaries, the systemic capillaries are modelled as a separate compartment in order to provide a more accurate description of the transcapillary fluid transport. The pulmonary capillaries were included in the pulmonary arterial compartment given a much lower difference in pressures between these two compartments [14] and the fact that the transcapillary fluid transport in the pulmonary bed is not modelled directly (see the discussion in Sect. 3.3).

All cardiovascular compartments are divided in the model in two subcompartments representing plasma and red blood cells (the latter occupying typically almost half of the vascular space on the whole-body level). Having separate RBC compartments enables accounting for a different chemical composition of plasma and RBC fluid as well as for differences in the permeability of the red cell membrane to different solutes (see Table 2.5 in Sect. 2.4.2), which affects the transport processes during HD. Moreover, separate RBC compartments enable the description of the different level of HCT between the micro and macro vessels (see Sect. 2.2.5) as well as the description of HCT changes during HD (both in individual vascular compartments and on the whole-body level), which in turn affects the vascular resistance (see Sect. 2.2.2). Finally, separate RBC compartments enable the description of possible osmotic water shifts between plasma and RBCs during HD affecting the total volume of RBCs and HCT. For simplicity, other blood components (such as white blood cells or platelets) were included in the plasma compartment given their rather negligible volume (<1%).

The compartment volumes (the sum of plasma and red blood cell volume) are model variables, whereas the transmural pressures are derived variables. In this study, linear pressure-volume relationships are used for all cardiovascular compartments, thus resigning from nonlinear *P-V* relationships for systemic veins and vena cava introduced in the VM model [2] (during HD venous pressure variations are much lower than during the VM, and hence a linear approximation of the pressure-volume relationship is sufficient for this study [39]):

$$P_i = \frac{(V_i - V_{u,i})}{C_i} \tag{2.4}$$

where i denotes the i-th compartment, P, transmural pressure; V, compartment volume; V_u, compartment unstressed volume; and C, compartment compliance (assumed constant).

The unstressed volumes of all compartments are calculated from the assumed normal pressures and volumes (see Tables 2.6 and 2.7) and the assumed vascular compliances (see Table 2.1). Since the real vessels feature nonlinear pressure-volume relationships [49, 50], the thusly calculated unstressed volumes are likely overestimated (see Fig. 2.2). This is, however, compensated by the fact that the nominal pressure used for the given compartment in the model represents the pressure at its entry and not the average pressure in the vessels.

All compliances are scaled according to the body weight of the given individual as follows:

$$C_i = C_{i,\mathrm{kg}} \cdot \mathrm{BW} \tag{2.5}$$

where $C_{i,\mathrm{kg}}$ is the compliance of the i-th compartment per kg of body weight (see the values assigned in Sect. 2.4.1), BW is patient's body weight.

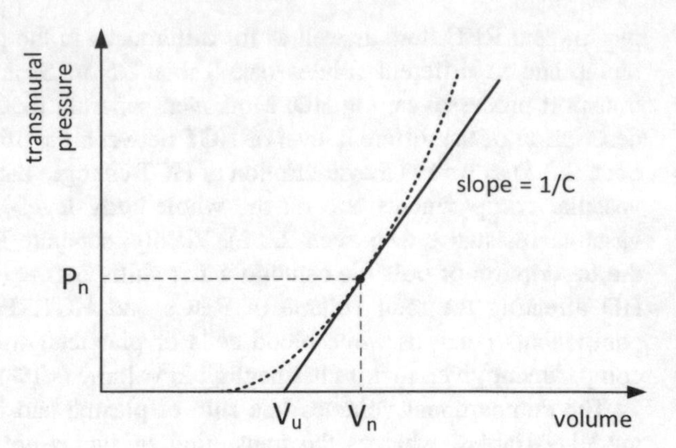

Fig. 2.2 The generic linear pressure-volume relationship (solid line) used in the model for all cardiovascular compartment vs the typical nonlinear pressure-volume relationship (dashed line) seen in real vessels [49, 50]. *C*, vascular compliance; P_n, normal transmural pressure; V_n, normal volume; V_u, unstressed volume

2.2.2 Vascular Resistance

Vascular resistance depends on the geometry of the vasculature and corresponds to the pressure losses due to friction between the blood and vessel walls (dependent, among others, on the vessel lumen and length) as well as to the local pressure losses related to the change of flow direction (e.g. bends, narrowings, bifurcations etc.) [2].

According to the Hagen-Poiseuille's law [51], for a laminar flow of a Newtonian fluid, the resistance of a straight long narrow rigid cylindrical tube can be expressed as follows:

$$R = \frac{8 \cdot \eta \cdot L}{\pi \cdot r^4} \tag{2.6}$$

where η is the fluid dynamic viscosity, L is the tube length and r is the tube radius.

Since blood is not a Newtonian fluid, the blood vessels are not rigid and do not have a uniform circular cross-section and the blood flow is not always laminar (especially in the large arteries), the Hagen-Poiseuille's law does not strictly apply to the flow of blood through the vascular system. Nevertheless, for simplicity, assuming a constant length and circular cross-section of the vessels represented by each of the modelled compartments, based on the above relationship, it was assumed that the vascular resistance is inversely proportional to the volume squared [50, 52]:

$$R \sim \frac{1}{r^4} \sim \frac{1}{V^2} \tag{2.7}$$

As far as the blood dynamic viscosity (η) is concerned, it was assumed that it depends on haematocrit (HCT), which can change as a result of blood volume changes during HD [53–55]. To account for this phenomenon, the following linear function was used to relate blood viscosity to plasma viscosity, which is a good approximation for low haematocrit values [56] (as seen in dialysis patients [57, 58]):

$$\eta = \eta_{\text{plasma}}(1 + 2.5 \cdot \text{HCT}) \tag{2.8}$$

Based on the above equations, the following relationship between the vascular resistance, volume and haematocrit was derived:

$$R_\text{i} = \frac{\kappa_\text{i}(1 + 2.5 \cdot \text{HCT}_\text{i})}{V_\text{i}^2} \tag{2.9}$$

where V_i and HCT_i are volume and haematocrit of the i-th vascular compartment and κ_i is a parameter calculated for the i-th compartment based on initial (normal) conditions (i.e. normal volume, resistance and haematocrit – see Sect. 2.5.1).

For simplicity, the above function does not include a minimal (basal) value of vascular resistance used by some authors [50, 52, 59]. With this approach, the resistance depends solely on the compartment volume and haematocrit, thus neglecting the impact of other geometrical determinants of vascular resistance, such as the curvature, branching or tapering of vessels, and hence the model slightly overestimates the resistance changes following changes in the compartment volume.

In the above approach, a constant blood temperature and a constant plasma viscosity were assumed, thus ignoring possible changes of these values during HD [53, 60]. The possible changes of blood viscosity due to changes in shear stress (related to blood velocity), protein concentration or white blood cells count were also neglected. The impact of vessel diameter on blood viscosity in the microcirculation (the Fåhraeus-Lindqvist effect [61, 62]) and the impact of the endothelial surface layer (also known as glycocalyx, with a diameter-dependent thickness) on blood viscosity [63] were ignored.

2.2.3 *Cardiac Output*

Despite the fact that the proposed model is a non-pulsatile model (the blood is ejected from the cardiac compartments continuously rather than on a stroke basis), the cardiac output of right and left heart compartments is described as a product of heart rate (HR) and stroke volume (SV) [4, 5]. For the left heart output we have:

$$q_\text{l} = \text{SV}_\text{l} \times \text{HR} \tag{2.10}$$

The stroke volume is a function of preload (i.e. the end-diastolic stretch of cardiac muscles typically represented by atrial pressure) and afterload (i.e. the force against which the heart must work during contraction) [14]. The relationship between SV and preload is usually modelled as either linear [4, 5] or exponential with upper saturation level [64, 65]. In this study, in order to better reflect the sigmoidal relationship between cardiac output and preload [14], the following sigmoid function was used, as introduced in [2]:

$$\mathrm{SV_l} = \frac{\mathrm{SV_{max}}}{1 + \exp\left(\frac{-P_{\mathrm{lat}} - x_l}{s_l}\right)} \cdot E \cdot a_l \tag{2.11}$$

where $\mathrm{SV_{max}}$ represents the maximal stroke volume, P_{lat} is the left atrial pressure (the pressure in the left heart compartment representing the cardiac preload), x_l and s_l are parameters of the stroke volume curve, E describes heart contractility controlled by the baroreflex (under normal conditions $E = 1$) and $\mathrm{a_l}$ describes the impact of afterload, as described below. It was assumed that the above relationship for SV includes implicitly all pressure losses occurring within the heart (e.g. across the valves).

The left ventricular afterload was approximated by the mean pressure in the large arteries, assuming that this pressure corresponds to the pressure generated by the left ventricle and hence describes the stress the ventricle is subject to (in a model with a pulsatile blood flow, the systolic pressure would be a better, albeit still not ideal, measure of afterload). As shown by several authors [66, 67], increasing afterload causes a reduction in the velocity and degree of myofibers shortening and hence a reduction in the ventricle performance. The inverse relationship between SV and afterload was shown by several authors both in isolated hearts and in vivo experiments, both with a constant and variable preload [68–70]. The impact of increased afterload was included in previous models of CVS [4, 5]. In this study the following function (proposed already for the VM model [2]) was used to describe the impact of both increased and reduced afterload on the left ventricular SV:

$$a_l = (1 + m) - m \cdot \left(\frac{P_{\mathrm{la}}}{P_{\mathrm{la,n}}}\right)^2 \tag{2.12}$$

where $P_{\mathrm{la,n}}$ is the normal mean arterial pressure and m is a parameter defining by how much the stroke volume can increase from the normal value with the afterload reduced to zero (in this study m was taken as 20% to match approximately the literature data [69–71].

The above approach to modelling the impact of afterload (shown graphically in Fig. 2.3a) provides the following features characteristic for the SV dependence on afterload (arterial pressure):

- The stroke volume increases moderately with a reduction in afterload [69].
- The changes in stroke volume are relatively small in the physiological range of aortic pressure [66, 72].
- The stroke volume is substantially reduced at higher afterload levels [68, 70].
- The relationship between the stroke volume and afterload is approximately linear in the physiological range [71, 73].

Moreover, the resulting description of the relationship between the afterload and stroke work (calculated here as a product of stroke volume and aortic pressure) takes the characteristic parabolic shape (see Fig. 2.3a) with the stroke work increasing up to a maximum level, beyond which it decreases with a further increase in afterload

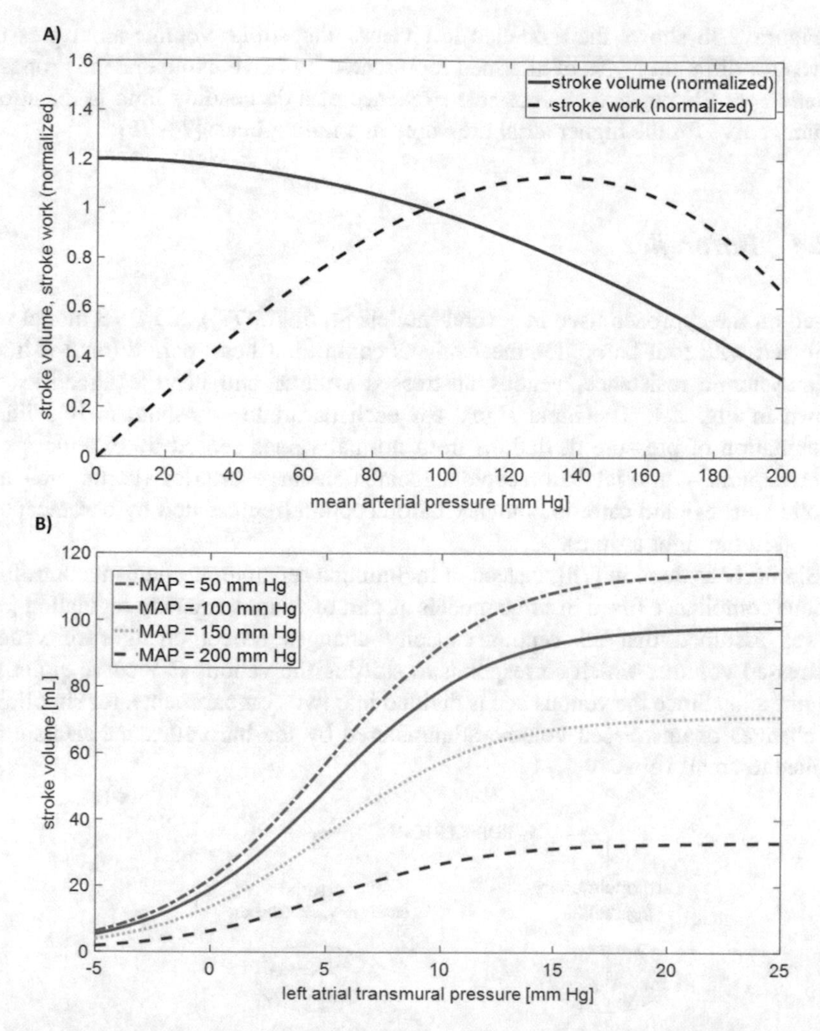

Fig. 2.3 (**a**) Simulated left ventricular stroke volume (solid line) and stroke work (dashed line) as a function of afterload (represented by MAP) normalised by the values obtained at normal MAP (95 mmHg); stroke work was calculated as a product of stroke volume and aortic pressure. (**b**) The impact of afterload (MAP) on left ventricular stroke volume-atrial pressure curve

[66, 67, 69]. Note that the location of the peak of the real stroke work curve with respect to the normal afterload (normal arterial pressure) may vary, with some experiments showing that a normal heart is operating somewhere near the maximum work level [68, 69].

Similar equations are used for the right ventricular stroke volume with the pulmonary arterial pressure instead of large arteries pressure used for calculating the impact of afterload.

Figure 2.3b shows the modelled left ventricular stroke volume-atrial pressure curves for different levels of afterload represented by MAP. Note that the proposed equation for SV neglects the possible existence of a descending limb in the stroke volume curve for the higher atrial pressures in a failing heart [74, 75].

2.2.4 *Baroreflex*

Based on the approach used in several models [4, 5, 76, 77]), the CVS model was equipped with four baroreflex mechanisms controlling heart period (or *R-R* interval), systemic resistance, venous unstressed volume and heart contractility, as shown in Fig. 2.4. The input signal for each baroreflex mechanism is a linear combination of pressure deviations from normal values sensed by two groups of baroreceptors – arterial baroreceptors located in large arteries (aortic arch and carotid sinuses) and cardiopulmonary baroreceptors (represented by baroreceptors located in the right atrium).

Similarly as done in [78], instead of including a separate mechanism controlling venous compliance (used in some models as part of venous capacity regulation [5]), it was assumed that all venous capacity changes reflect changes of venous unstressed volume, which corresponds to shifting the venous *P-V* curve along the volume axis. Since the venous bed is divided into two compartments, for simplicity, all changes of unstressed volume administered by the baroreflex mechanism are applied to small veins.

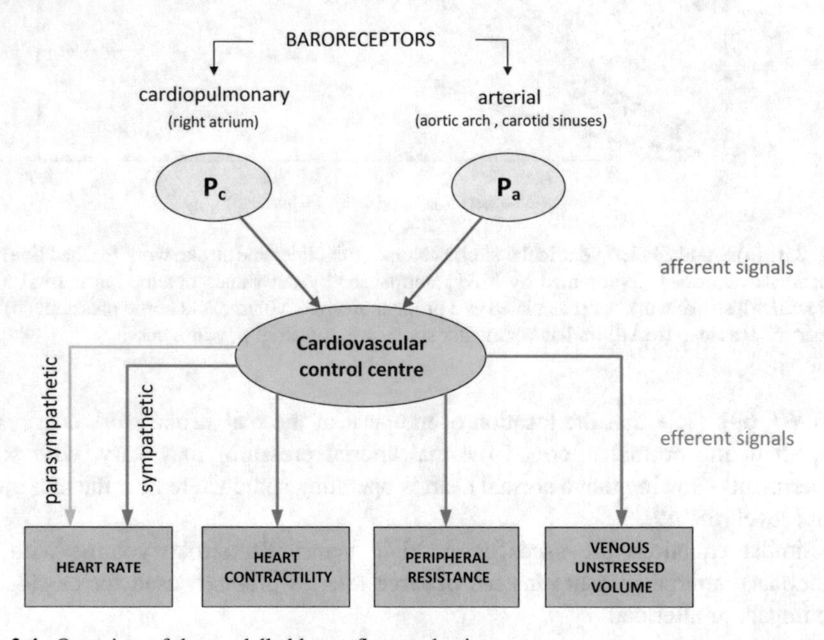

Fig. 2.4 Overview of the modelled baroreflex mechanisms

The control of systemic resistance (which in reality corresponds mainly to changes in arterioles resistance) is typically modelled explicitly [4, 5, 77]. In this study, because of the assumed relationship between the vessel volume and its resistance, this was not straightforward, since the volume-related changes in arterial resistance would affect the baroreflex effector response. Therefore, similarly to the approach proposed in the VM model [2], the resistance of small arteries and arterioles compartment was divided into two components – the resistance of small arteries (related to the volume of the compartment) and the resistance of arterioles controlled by the baroreflex mechanism (not involving any volume changes). The resistance of the small arteries compartment, affecting the blood flow from small arteries to capillaries, can therefore change in the model either due to changes in the volume of this compartment or (mainly) due to baroreflex operation (or as an effect of both mechanisms).

As is well known, changes of heart rate influence the time available for cardiac filling which affects the stroke volume (with a higher heart rate there is less time for the heart to fill up with blood, which decreases the end-diastolic volume and hence reduces the stroke volume). Given the relatively low heart rate changes during HD [79], this phenomenon was neglected in this study, and it was assumed that heart rate variations do not affect the stroke volume-atrial pressure relationship. The latter is, however, controlled by the baroreflex-induced changes in heart contractility through varying parameter E in the Eq. (2.11) (the higher the contractility, the higher the stroke volume).

For each baroreflex mechanism, similarly as done in [5] and [4], a sigmoidal function (with upper and lower saturation levels) was used to model the baroreflex response, with linear first-order dynamics to describe the ultimate effector responses (see Fig. 2.5).

For the control of systemic arterioles resistance (R) we have [4, 5]:

$$\frac{\mathrm{d}R}{\mathrm{d}t} = \frac{\sigma_{\mathrm{R}} - R}{\tau_{\mathrm{R}}} \tag{2.13}$$

where τ_{R} represents the time constant of the feedback mechanism and $\sigma_{\mathrm{R}}(X_{\mathrm{R}})$ is the sigmoidal static characteristic defined as [4]:

$$\sigma_{\mathrm{R}} = \frac{R_{\min} + R_{\max} \cdot \exp\left(\frac{-X_{\mathrm{R}}}{k_{\mathrm{R}}}\right)}{1 + \exp\left(\frac{-X_{\mathrm{R}}}{k_{\mathrm{R}}}\right)} \tag{2.14}$$

where $R_{\min}$ and $R_{\max}$ are minimal and maximal resistance values, k_{R} is a parameter that determines the slope at the central point of the sigmoidal function σ_{R} and X_{R} is an algebraic weighted sum of pressure deviations at the level of arterial and cardio-pulmonary baroreceptors [4]:

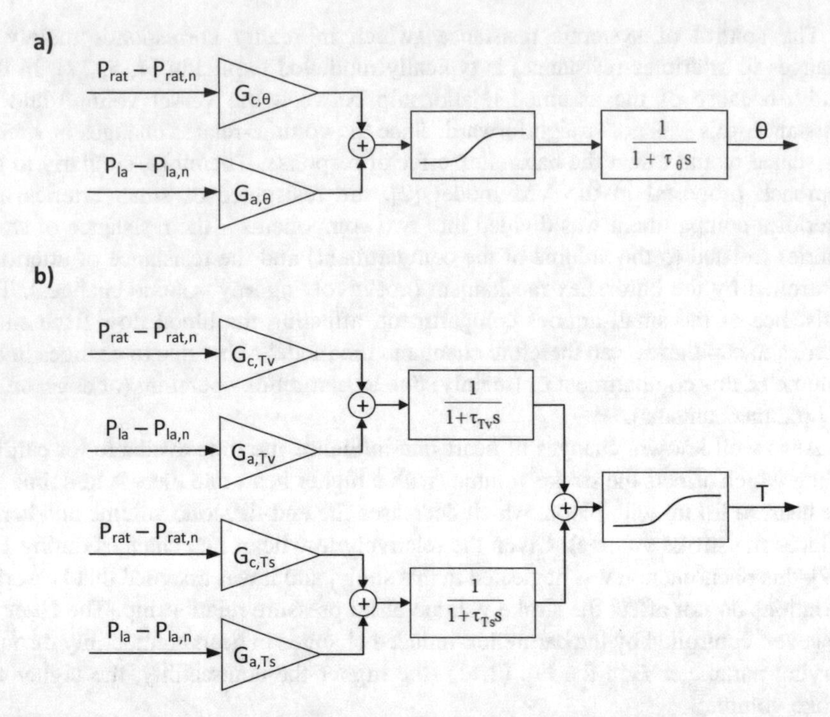

Fig. 2.5 Block diagram describing (**a**) sympathetically mediated baroreflex mechanisms where θ represents either resistance of systemic arterioles (R), unstressed volume of systemic veins ($V_{u,sv}$) or heart contractility (E); (**b**) baroreflex mechanism working on heart period (T) based on both sympathetic (s) and vagal (v) activity (Adapted from [77]). G_c and G_a represent cardiopulmonary and arterial gains (weights) of each baroreflex mechanism

$$X_R = G_{\mathrm{c,R}} \cdot (P_{\mathrm{rat}} - P_{\mathrm{rat,n}}) + G_{\mathrm{a,R}} \cdot (P_{\mathrm{la}} - P_{\mathrm{la,n}}) \tag{2.15}$$

where $G_{\mathrm{c,R}}$ and $G_{\mathrm{a,R}}$ represent the cardiopulmonary and arterial gains (weights) of the mechanism controlling peripheral resistance, P_{rat} and $P_{\mathrm{rat,n}}$ are the current and normal right atrial pressure and P_{la} and $P_{\mathrm{la,n}}$ are the current and normal pressure in the large arteries.

Similar equations are used for the control of systemic veins unstressed volume and heart contractility.

For the control of venous unstressed volume of small veins [4, 5]:

$$\frac{\mathrm{d}V_{\mathrm{u,sv}}}{\mathrm{d}t} = \frac{\sigma_{\mathrm{V}} - V_{\mathrm{u,sv}}}{\tau_{\mathrm{V}}} \tag{2.16}$$

$$\sigma_{\mathrm{V}} = \frac{V_{\mathrm{u,sv,min}} + V_{\mathrm{u,sv,max}} \cdot \exp\left(\frac{X_{\mathrm{V}}}{k_{\mathrm{V}}}\right)}{1 + \exp\left(\frac{X_{\mathrm{V}}}{k_{\mathrm{V}}}\right)} \tag{2.17}$$

$$X_V = G_{c,V} \cdot (P_{rat} - P_{rat,n}) + G_{a,V} \cdot (P_{la} - P_{la,n}) \tag{2.18}$$

where $V_{u,sv,min}$ and $V_{u,sv,max}$ are minimal and maximal unstressed volume of small veins, k_V is a parameter that determines the slope at the central point of the sigmoidal function σ_V, $G_{c,V}$ and $G_{a,V}$ represent the cardiopulmonary and arterial gains (weights) of the mechanism controlling venous unstressed volume, P_{rat} and $P_{rat,n}$ are the current and normal right atrial pressure and P_{la} and $P_{la,n}$ are the current and normal pressure in the large arteries.

For the control of heart contractility [4, 5]:

$$\frac{dE}{dt} = \frac{\sigma_E - E}{\tau_E} \tag{2.19}$$

$$\sigma_E = \frac{E_{min} + E_{max} \cdot \exp\left(\frac{-X_E}{k_E}\right)}{1 + \exp\left(\frac{-X_E}{k_E}\right)} \tag{2.20}$$

$$X_E = G_{c,E} \cdot (P_{rat} - P_{rat,n}) + G_{a,E} \cdot (P_{la} - P_{la,n}) \tag{2.21}$$

where E_{min} and E_{max} are minimal and maximal cardiac contractility, k_E is a parameter that determines the slope at the central point of the sigmoidal function σ_E, $G_{c,E}$ and $G_{a,E}$ represent the cardiopulmonary and arterial gains (weights) of the mechanism controlling cardiac contractility, P_{rat} and $P_{rat,n}$ are the current and normal right atrial pressure and P_{la} and $P_{la,n}$ are the current and normal pressure in the large arteries.

For the regulation of heart period (*R*-*R* interval), which is based on both sympathetic and vagal activity, separate gains and distinct time constants were used for both submechanisms [77].

$$T = \frac{T_{min} + T_{max} \cdot \exp\left(\frac{X_T}{k_T}\right)}{1 + \exp\left(\frac{X_T}{k_T}\right)} \tag{2.22}$$

$$X_T = X_{Tv} + X_{Ts} \tag{2.23}$$

$$\frac{dX_{Tv}}{dt} = \frac{\delta_{Tv} - X_{Tv}}{\tau_{Tv}} \tag{2.24}$$

$$\frac{dX_{Ts}}{dt} = \frac{\delta_{Ts} - X_{Ts}}{\tau_{Ts}} \tag{2.25}$$

$$\delta_{Tv} = G_{c,Tv} \cdot (P_{rat} - P_{rat,n}) + G_{a,Tv} \cdot (P_{la} - P_{la,n}) \tag{2.26}$$

$$\delta_{Ts} = G_{c,Ts} \cdot (P_{rat} - P_{rat,n}) + G_{a,Ts} \cdot (P_{la} - P_{la,n}) \tag{2.27}$$

where T_{min} and T_{max} are minimal and maximal heart period, k_T is a parameter that determines the slope at the central point of the sigmoidal function T, G_c and G_a

represent the cardiopulmonary and arterial gains (weights) of the vagal (T_v) and sympathetic (T_s) mechanisms controlling heart period, P_{rat} and $P_{rat,n}$ are the current and normal right atrial pressure and P_{la} and $P_{la,n}$ are the current and normal pressure in the large arteries.

Depending on the sign before the input signal (X), the above sigmoidal relationships (Eqs. 2.14, 2.17, 2.20, and 2.22) are either monotonically increasing or decreasing. For the control of systemic resistance, the negative sign is used before X_R in Eq. (2.14) to describe the inverse relationship between the blood pressure and systemic resistance (when the blood pressure is too high, the resistance should be lowered and vice versa). Similarly, the negative sign is used before X_E in Eq. (2.20). In the case of the venous capacity regulation, the relationship between blood pressure and venous unstressed volume is positive (when the blood pressure is too high, the venous unstressed volume should be increased to reduce the venous pressure and cardiac filling), hence the positive sign before X_V in Eq. (2.17). Similarly, the positive sign is used before X_T in Eq. (2.22). For the values of parameters used in the above equations, please see Sect. 2.4.1.

Note that looking at the individual baroreflex mechanisms separately (with other mechanisms excluded from the system), the role of the low-pressure (cardiopulmonary) and high-pressure (arterial) baroreceptors may seem antagonistic in some circumstances [2]. For instance, with a normal arterial pressure and an elevated right atrial pressure (or central venous pressure), according to the Eqs. (2.13, 2.14, and 2.15), the resistance of arterioles should be lowered by the baroreflex mechanism which would decrease the arterial pressure while increasing even more the venous and atrial pressure. This is, of course, related to the closed-loop nature of the cardiovascular system in which for the arterial pressure to decrease, some blood must be transferred from the arteries to the venous part of the system. One must remember that all four baroreflex mechanisms have direct or indirect effects on both arterial and venous side of the system. Only the concomitant action of all mechanisms (combined with the regulation of blood volume through transcapillary fluid shifts, described later) can provide a correct control of the whole cardiovascular system.

2.2.5 *Haematocrit*

Haematocrit (HCT), i.e. the ratio of RBC volume to the whole blood volume, is not constant across the CVS [15]. In the microcirculation (i.e. arterioles, capillaries and venules), due to low vessel diameter (<500 μm), HCT is much lower than in the macrocirculation (the Fåhraeus effect) [80, 81]. Therefore, the whole-body HCT (HCT_{body}) is lower than the central HCT [82–84]. This is related to the fact that in an average microvessel, there is a fast-moving axial flow of RBCs surrounded by a cell-free marginal layer (several microns thick) of slowly moving plasma [84], as explained by the marginal zone theory proposed by Haynes [85], and hence the HCT of blood flowing out of the vessel (the so-called discharge haematocrit, HCT_D) is higher than the average HCT of the total blood within the given vessel (the so-called tube haematocrit, HCT_T) [15, 81, 86]. For the macrocirculation $HCT_D = HCT_T$ [15].

The whole-body HCT can be calculated as follows [87]:

$$HCT_{body} = \frac{\sum_i HCT_{T,i} V_i}{TBV} \tag{2.28}$$

where $HCT_{T,i}$ is the tube haematocrit in the i-th blood compartment, V_i is the volume of the i-th blood compartment and TBV is the total blood volume.

Based on the assumed normal central HCT assigned to the large arteries, large veins and cardiac compartments (HCT_C) and the assumed ratio of whole-body HCT to central HCT (also known as the F-cells ratio [82]), a low-level tube HCT was assigned to the systemic capillaries compartment (see the initial conditions described in Sect. 2.5.1). For all other vascular compartments containing both vessels with normal and low HCT (i.e. small arteries, small veins, pulmonary arteries and pulmonary veins), a medium-level HCT was assigned based on the ratio of high-HCT blood volume and low-HCT blood volume (see Sect. 2.5.1). It was assumed that arterioles and venules have a similarly low haematocrit, thus neglecting the fact that the blood velocity in venules is slightly lower than in arterioles, and hence the HCT is slightly higher in venules than in arterioles [6, 7]. For simplicity, the smallest capillaries (diameter < 4 μm), in which RBCs squeeze through without leaving space for the marginal layer of plasma [8, 62], were ignored.

Once the initial HCT values are assigned for all compartments (see Sect. 2.5.1), one can distinguish between the plasma volume and RBC volume in each blood compartment. In order for the haematocrit not to equalise across the whole CVS at the steady-state conditions, the outflow of RBCs from each compartment must reflect the actual discharge haematocrit of the given compartment.

For macrocirculation compartments (i.e. large arteries, large veins and cardiac compartments) $HCT_D = HCT_T$, i.e. the outflow rate of RBCs from the i-th compartment ($Q_{rc,out,i}$) is equal to the product of blood outflow ($Q_{b,out,i}$) and the tube HCT of the given compartment (HCT_i):

$$Q_{rc,out,i} = HCT_{T,i} \cdot Q_{b,out,i} \tag{2.29}$$

For all other compartments (i.e. small arteries, capillaries, small veins, pulmonary arteries and pulmonary veins) $HCT_D > HCT_T$, and hence an additional coefficient was introduced to reflect the high apparent HCT of blood flowing out of these compartments:

$$Q_{rc,out,j} = \gamma_j \cdot HCT_{T,j} \cdot Q_{b,out,j} \tag{2.30}$$

To ensure initial steady-state conditions (i.e. no accumulation or depletion of RBCs in any compartment), the coefficient γ_j is defined as follows (see Fig. 2.6):

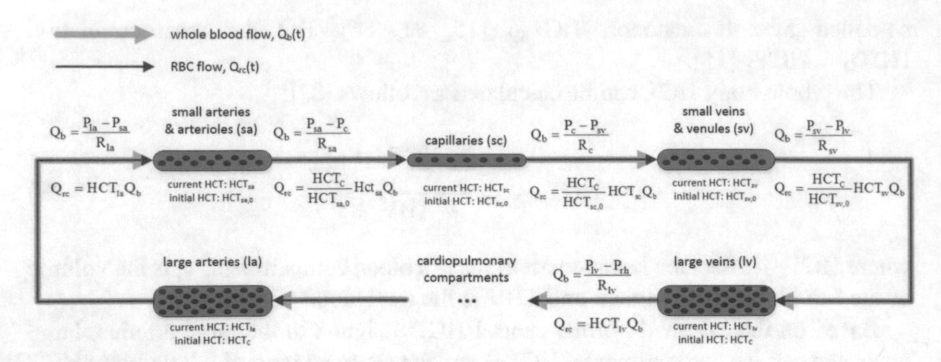

Fig. 2.6 Model representation of plasma and red blood cells flow across the cardiovascular system

$$\gamma_j = \frac{HCT_C}{HCT_{T,j,0}} \tag{2.31}$$

where HCT_C is the normal haematocrit assumed for the central compartments, $HCT_{T,j,0}$ is the initial tube HCT assumed for the *j*-th compartment.

For calculating the parameters γ_j, it was assumed that there is no fluid filtration at the capillary level, given that normal filtration is much lower than the blood flow rate [14]. For simplicity, it was assumed that the values of parameters γ_j remain constant throughout haemodialysis (more detailed relationships between the microvessel tube HCT and the discharge HCT can be found in the literature [8, 81]).

Following the above correction of the RBC outflows of microcirculation compartments, the plasma outflows must be also modified accordingly, so that in equilibrium the whole blood flow is constant. The latter is always calculated from the pressure difference between the adjacent compartments (see Eq. 2.2).

The quantitative outflow of RBCs from each compartment is calculated by dividing the volumetric outflow by the mean corpuscular volume (MCV) of RBCs in the given compartment (see Eq. 2.45). This enables keeping track of the number of RBCs in each compartment, which itself is used for calculating the instantaneous MCV of RBCs in each blood compartment and the local exchange surface area between plasma and RBC fluid needed for the calculations of transport of solutes and water across the RBC membrane. The initial number of RBCs in each compartment was calculated assuming the average MCV equal 90 μm^3 [9] (it was assumed that RBCs have the same volume across the whole CVS, thus neglecting the possible differences in MCV between capillary and venous blood [10]). A constant number of RBCs was assumed for the whole system, i.e. the natural turnover of RBCs and the possible damage of RBCs in the extracorporeal circuit were ignored, both factors having relatively little importance during a single few-hour HD [11].

2.2.6 Arteriovenous Access

As already mentioned in Sect. 2.2.1, the proposed model includes an arteriovenous fistula (or graft) as a standard vascular access for HD. It was assumed that the fistula connects brachial artery with cephalic vein in the lower arm, and therefore it is placed between large arteries and large veins compartment – see Fig. 2.1 in Sect. 2.2.1 and Fig. 2.7 in Sect. 2.3.1. It was assumed that the fistula is connected to the end of large arteries and to the beginning of large veins, and hence the blood flowing through the fistula bypasses the microcirculation. The volume of fistula was neglected (it was treated purely as a flow bypass). Due to the lumped parameter nature of the CVS model, the possible inter-arm differences in blood pressure or blood flow due to the presence of fistula are neglected.

The access blood flow rate is calculated from the following equation:

$$Q_{ac} = \frac{P_{sa} - P_{lv}}{R_{ac}} \tag{2.32}$$

where R_{ac} is the access resistance (assumed constant and calculated from the pre-dialysis steady-state conditions, see Sect. 2.5.2).

The extracorporeal circuit consisting of three blood compartments – arterial tubing (V_{at}), dialyzer (V_d) and venous tubing (V_{vt}) – is connected directly to the vascular access (see Fig. 2.7 in Sect. 2.3.1). The model includes three extracorporeal compartments in order to better describe the blood flowing through the dialyzer (with seperate compartments describing blood inflow and outflow to/from the dialyzer) as well as to enable simulation of fluid infusions during HD or other dialysis techniqes (such as haemofiltration). The volume of the extracorporeal circuit was assumed to be constant (rigid tubing), with the blood flow to the dialyzer determined by the pump (Q_d) and the blood flow out of the dialyzer being the difference between the blood inflow and dialyzer ultrafiltration (Q_{uf}). A healthy and well-functioning fistula was assumed with correctly placed needles, and hence no access recirculation was considered (note that the cardiopulmonary recirculation [88] is included in the model explicitly). It was also assumed that the operation of the artificial kidney machine pump does not alter the blood flow entering the arteriovenous access [89].

If needed, the model can be modified to describe a central veno-venous dialysis access (a permanent catheter) by connecting both arterial and venous dialyzer tubing compartments to the large venous compartment.

2.2.7 Assumptions

Apart from several model assumptions mentioned or discussed earlier, the following simplifications were assumed for the modelled cardiovascular system:

1. The blood is an incompressible and Newtonian fluid.
2. Active response of vascular smooth muscles to blood pressure changes is negligible.
3. Vascular viscoelastic effects (stress relaxation) are negligible.
4. There is no hysteresis in the vascular pressure-volume curves.
5. The pressure waves reflected from vessel bifurcations are negligible.
6. The body is in the supine position.
7. The effects of muscle pump on venous return are negligible.
8. The effects of respiration on heart rate variations are negligible.
9. The Anrep effect (an increase in heart contractility at increased afterload [90]) is negligible.
10. There is no pressure "talk" between right and left cardiac chambers (as modelled in [91]).
11. There is no time latency in baroreflex operation.
12. Baroreceptors are not sensitive to the rate of pressure changes.
13. There are no other mechanisms controlling blood pressure (e.g. chemoreceptors).
14. There is no regional blood flow autoregulation (e.g. in brain, heart or kidneys).

The above assumptions demonstrate the sophistication and complexity of the human cardiovascular system and the blood pressure regulatory mechanisms. The details of the phenomena neglected by the above assumptions are beyond the scope of this study. For a brief discussion on these and other limitations of the developed model, see Sect. 3.3.

2.3 Water and Solute Kinetics

2.3.1 Fluid Compartments

The following fluid compartments are considered in the model (see Fig. 2.7): interstitial fluid (is), extravascular intracellular fluid (ic), blood plasma (pl) and red blood cells fluid (rc). The last two intravascular compartments are additionally divided among all blood compartments (i.e. for each blood compartment, there is a separate plasma compartment and a separate RBC compartment). For simplicity, the interstitial fluid represents all extravascular extracellular fluids including the interstitial fluid of muscles, skin, brain and viscera (both free fluid and non-mobile gel fluid), fluid in the bones (including the extracellular marrow fluid) and dense connective tissues as well as transcellular fluids (i.e. synovial, peritoneal, pericardial, pleural, ocular or cerebrospinal fluids, gastrointestinal duct secretions, sweat, tears and other body fluids) [14, 92]. Due to the low volume of the lymphatic system, the lymph was treated as part of the interstitial fluid [14, 92]. The extravascular intracellular fluid represents fluid within the tissue cells, which were assumed to be represented by the skeletal muscle cells, given that the mass of skeletal muscle tissue

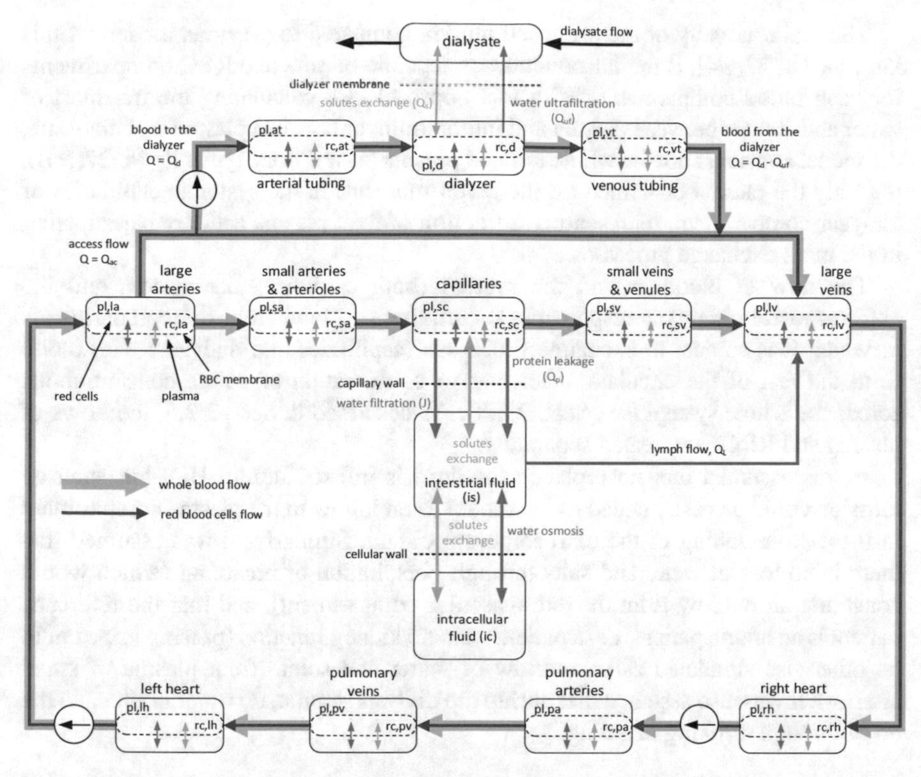

Fig. 2.7 Schematic representation of water and solute exchange across the cardiovascular and extracorporeal blood compartments during haemodialysis

is the largest in the body [14], it is associated with 65% of the total microvascular exchange [12], and over 80% of the extravascular intracellular fluid is the fluid within the skeletal muscle cells [93].

All fluid compartments are assumed to be homogeneous (i.e. well mixed with spatially constant parameters at any given time) and represent the lumped properties of the given space. In particular, it was assumed that all components of the extravascular extracellular space are in equilibrium at all times, and hence the possible time delays in water and solute transfer between the interstitial fluid of muscles or organs and less accessible extracellular water (e.g. water in the bones, dense connective tissues or transcellular spaces) were ignored. The blood-brain barrier was also ignored.

With the above approach, the model accounts for the total body water (except water bound in the minerals), and hence it describes the whole-body water and solute kinetics. The volumes and composition of all fluid compartments change according to the shifts of water and solutes between the compartments, i.e. across the capillary, cellular and dialyzer membranes, as well as via the lymphatic system. The initial volumes of all compartments were assigned based on the literature data (see Sect. 2.5).

The major novelty of the proposed model, compared to previous models of this kind [4, 18, 27, 94], is the introduction of separate plasma and RBC compartments for each blood compartment. With this approach, for calculating the transport of water and solutes between plasma and interstitium, or between plasma and dialysate, the model considers not the whole lumped plasma (as it is done typically [4, 27, 94]), but only the plasma contained (at the given moment) in the systemic capillaries or dialyzer compartment, representing a fraction of total plasma actually participating in the mass exchange processes.

The flow of blood around the system (both cardiovascular system and the extracorporeal circuit) is responsible for mixing of blood from the compartments in which it is subject to exchange processes (capillaries and dialyzer) with blood from the rest of the circulation, leading to equilibration of solute concentrations across the whole system (see Sect. 2.3.8). As described in Sect. 2.2.5, the flows of plasma and RBCs are treated separately.

It was assumed that no replacement fluid is infused during HD, but such an infusion could be easily added to the model as an inflow to the plasma compartment in the venous tubing of the extracorporeal circuit. Similarly, it was assumed that there is no loss of water and salts through perspiration or breathing (which would constitute an outflow from the extravascular compartment), and that the reference patient is an anuric patient, i.e. a patient with no kidney function (passing urine could be otherwise simulated as an outflow of water and solute from plasma of small arteries). It was also assumed that during the dialysis session, no water is added to the body through drinking or eating.

2.3.2 *Small Solute Kinetics*

The model describes the whole-body distribution and transport of the following ions and small, low-weight molecules: sodium (Na^+), potassium (K^+), chloride (Cl^-), bicarbonate (HCO_3^-), urea (U) and creatinine (Cr). Compared to the Ursino and Innocenti model [4], a few more solutes were included in the model (Cl^-, HCO_3^-, Cr) in order to describe more accurately the fluid-electrolyte balance and changes in intracellular and extracellular osmolarity, as well as to enable a more detailed analysis of dialysis adequacy, which is typically focused on assessing urea and creatinine kinetics based on simple models [95, 96].

Similarly to the model by Gyenge et al. [27], the model accounts for other osmotically active cations and anions, which are described collectively as Cat^{2+} and An^{2-}. A charge of +2 was assigned to cations assuming that these represent mainly Mg^{2+} and Ca^{2+}. Anions were given an average charge of −2 given that they represent both monovalent and multivalent anions (e.g. $H_2PO_4^-$, HPO_4^{2-}, PO_4^{3-}, SO_4^{2-}).

Transport of solutes (with the generic solute denoted by s) takes place (1) between the intracellular and interstitial space (across the cellular membrane), (2) between the interstitial space and capillary plasma (across the capillary walls), (3) between blood plasma and RBCs in each blood compartment (across the RBC membrane), (4) between adjacent blood compartments (through blood flow), (5) between plasma and dialysate fluid (across the dialyzer membrane) and (6) between the interstitium and plasma of large veins (through lymph flow).

Concentration of solute s in the given compartment is calculated as the ratio of the number of moles of solute s and the volume of the compartment (molar concentration).

The model accounts for the generation of urea ($g_u = 310$ mmol/24 hours [20]), which was assigned to the large veins compartment (to which urea is delivered from the liver, where it is produced), and the generation of creatinine ($g_{Cr} = 9.8$ mmol/24 hours [20]), which was assigned to the intracellular fluid (assuming that it represents fluid of skeletal muscle cells, where creatinine is produced). For all other solutes, it was assumed that during HD their rate of intake or generation is negligible compared to the mass of solute exchanged in the dialyzer.

2.3.2.1 Cellular Membrane

As far as transport across tissue cell membrane is concerned, both passive diffusion and active transport mechanisms are considered for electrolytes (e.g. sodium-potassium pump).

For sodium, potassium, chloride and bicarbonate ($s = Na^+, K^+, Cl^-, HCO_3^-$), the diffusive transport from tissue cells to interstitium is given by the following relationship [4, 19]:

$$Q_{s,cell} = K_{s,cell}\left(\frac{c_{s,ic}}{F_{ic}} - \beta_{s,cell}\frac{c_{s,is}}{F_{is}}\right) \tag{2.33}$$

where $K_{s,cell}$ is the whole-body transcellular mass transfer coefficient for solute s; $c_{s,ic}$ and $c_{s,is}$ are intracellular and interstitial solute concentrations, respectively; F_{ic} and F_{is} are intracellular and interstitial fluid water fractions (variable); and $\beta_{s,cell}$ is a coefficient describing the ratio of intracellular and extracellular concentrations of solute s at equilibrium (due to active transport mechanisms).

For sodium, chloride and bicarbonate, which have lower concentrations inside the cells than outside, β is lower than 1, while for potassium, which is transported from interstitial fluid to the cells, β is greater than 1 [4].

Given that the changes of intracellular volume are very small during dialysis [97, 98], it was assumed that the cellular area of mass exchange is constant, and hence it was assumed that the mass transfer coefficients remain constant throughout dialysis.

For urea and creatinine (s = U, Cr), the convective transport was also included, thus modifying the above equation as follows [99, 100]:

$$Q_{s,cell} = K_{s,cell}\left(\frac{c_{s,ic}}{F_{ic}} - \frac{c_{s,is}}{F_{is}}\right) + J_{cell}(1 - \sigma_{s,cell})\left[(1-f)\frac{c_{s,ic}}{F_{ic}} + f\frac{c_{s,is}}{F_{is}}\right] \quad (2.34)$$

where J_{cell} is the osmotic water flow out of the cells (see Eq. 2.65), $\sigma_{s,cell}$ is the Staverman's reflection coefficient of solute s at the cellular membrane and f is defined as [99]:

$$f = \frac{1}{Pe} - \frac{1}{\exp(Pe) - 1} \quad (2.35)$$

where Pe is the Peclet number describing the relationship between the convective and diffusive transport [99] (if diffusion and convection are of opposite directions, the below equation takes the negative sign):

$$Pe = \frac{J_{cell}(1 - \sigma_{s,cell})}{K_{s,cell}} \quad (2.36)$$

For simplicity, it was assumed that other cations (Cat^{2+}) do not cross the cellular membrane (hence, their intracellular concentration changes only with changes of the cellular volume), while the flow of other anions (An^{2-}) is calculated to obtain a zero net flow of charge (z) across the cellular membrane as follows [27]:

$$\frac{dM_{An,ic}}{dt}z_A = -\frac{dM_{Na,ic}}{dt}z_{Na} - \frac{dM_{K,ic}}{dt}z_K - \frac{dM_{Cl,ic}}{dt}z_{Cl} - \frac{dM_{HCO_3,ic}}{dt}z_{HCO3} \quad (2.37)$$

2.3.2.2 Capillary Wall

The transcapillary transport of small solutes from the capillary plasma to interstitium (except other anions) was described using a similar equation as for the diffusive-convective transport across the cellular membrane (Eq. 2.34) but considering the Gibbs-Donnan effect for ions, i.e. accounting for the interactions between ions and negatively charged proteins [14, 15, 101, 102]:

$$Q_{s,cap} = p_{s,cap} \cdot S_{cap}\left(\alpha_{s,cap}\frac{c_{s,pl,sc}}{F_{pl,sc}} - \frac{c_{s,is}}{F_{is}}\right) + J_{cap}(1 - \sigma_{s,cap})\left[(1-f)\alpha_{s,cap}\frac{c_{s,pl,sc}}{F_{pl,sc}} + f\frac{c_{s,is}}{F_{is}}\right] \quad (2.38)$$

where $p_{s,cap}$ is the solute permeability of the lumped capillary wall, S_{cap} is the total capillary surface area (assumed constant), $\alpha_{s,cap}$ is the Gibbs-Donnan coefficient for ion s with charge z_s (for simplicity $\alpha_{s,cap} = \alpha_{cap}{}^{z_s}$ [103], where α_{cap} is determined from the steady-state conditions, see Sect. 2.5.1), $c_{s,pl,sc}$ is the concentration of solute s in the capillary plasma, $F_{pl,c}$ is plasma water fraction in the capillary compartment (variable), J_{cap} is water filtration from capillaries to interstitium (see Eq. 2.48), $\sigma_{s,cap}$ is the reflection coefficient of solute s at the capillary wall and f is defined similarly as in Eq. (2.35).

Given that protein concentration is larger in plasma than in the interstitium, it was assumed that the net Gibbs-Donnan effect takes place on the plasma side [14] with the Gibbs-Donnan coefficient α_{cap} calculated from the steady-state conditions (see Sect. 2.5.1).

Similarly as for the cellular membrane, the transport of other anions (An^{2-}) is given by electrical charge neutrality.

The permeability (in cm/s) of capillary wall to small solutes (urea and creatinine) was calculated as a function of the solute molecular weight (MW, g/mol) [24, 25]:

$$p_{s,cap} = 296 \cdot MW^{-0.63} \times 10^{-6} \tag{2.39}$$

The permeability of capillary wall to ions was assumed to be equal to urea permeability [104].

The changes of mass of solute s in the interstitial fluid are given by the following relationship [27, 105]:

$$\frac{dM_{s,is}}{dt} = Q_{s,cell} + Q_{s,cap} - c_{s,is}Q_L \tag{2.40}$$

where Q_L is the flow of lymph absorbed from the interstitium (see Eqs. 2.63 and 2.64).

2.3.2.3 Red Blood Cell Membrane

Small solutes transport across the RBC membrane in the i-th blood compartment can be expressed as follows (c.f. Eq. 2.33):

$$Q_{s,rc,i} = p_{s,rc}S_{s,rc,i}\left(\frac{c_{s,rc,i}}{F_{rc,i}} - \beta_{s,rc}\frac{c_{s,pl,i}}{F_{pl,i}}\right) \tag{2.41}$$

where $p_{s,rc}$ is the RBC membrane permeability to solute s, $S_{rc,i}$ is the surface area of RBCs in the i-th compartment (variable), $\beta_{s,rc}$ describes the equilibrium concentration ratio, $c_{s,pl,i}$ and $c_{s,rc,i}$ are the concentrations of solute s in plasma and RBCs of the i-th compartment and $F_{rc,i}$ is the RBC water fraction in the i-th compartment (variable).

Similarly as for the tissue cells, the transport of urea and creatinine (s = U, Cr) includes additionally the convective term (as in Eq. 2.34); transport of other anions is governed by electroneutrality (as in Eq. 2.37); it is assumed that other cations do not cross the RBC membrane. Note that the parameters describing the properties of cellular membrane are different for tissue cells and RBCs [31] (see Table 2.5 in Sect. 2.4.2).

The total surface area of RBCs is calculated separately for each compartment as follows:

$$S_{\mathrm{rc},i} = n_{\mathrm{rc},i} \cdot A_{\mathrm{rc},i} \tag{2.42}$$

where $n_{\mathrm{rc},i}$ is the number of RBCs in the i-th compartment and $A_{\mathrm{rc},i}$ is the surface area of a single RBC in the i-th compartment.

For simplicity, a cylindrical shape of RBCs was assumed, and hence:

$$A_{\mathrm{rc},i} = \pi d\left(\frac{d}{2} + t_i\right) \tag{2.43}$$

where d is the diameter of RBC (assumed to remain constant) and t_i is the thickness of RBC in the i-th compartment calculated as:

$$t_i = \frac{4\mathrm{MCV}_i}{\pi d^2} \tag{2.44}$$

where MCV_i is the mean corpuscular volume of a single RBC in the i-th compartment calculated as the total RBC volume in the given compartment ($V_{\mathrm{rc},i}$) divided by the total number of RBCs in the given compartment ($n_{\mathrm{rc},i}$):

$$\mathrm{MCV}_i = \frac{V_{\mathrm{rc},i}}{n_{\mathrm{rc},i}} \tag{2.45}$$

For calculating d a normal MCV of 90 μm^3 and $t = 2$ μm were assumed [14].

The above approach accounts for the fact that the mean volume (MCV) and surface (A_{rc}) may be slightly different for different vascular compartments given the intense solute and water transport processes occurring during HD, especially in the capillaries and dialyzer compartments.

The mass of solute s in the capillary plasma compartment changes at the rate described as:

$$\frac{\mathrm{d}M_{\mathrm{s,pl,sc}}}{\mathrm{d}t} = -Q_{\mathrm{s,cap}} + Q_{\mathrm{s,rc,sc}} + c_{\mathrm{s,pl,sa}}Q_{\mathrm{pl,in,sc}} - c_{\mathrm{s,pl,sc}}Q_{\mathrm{pl,out,sc}} \tag{2.46}$$

where $Q_{\mathrm{s,cap}}$ is the diffusive-convective transport of solute s as across the capillary walls (Eq. 2.38), $Q_{\mathrm{s,rc,sc}}$ is the transport of solute s across the RBC membrane in the capillary compartment (Eq. 2.41), $c_{\mathrm{s,pl,sa}}$ is the concentration of solute s in plasma of small arteries (supplying the capillaries), $Q_{\mathrm{pl,in,sc}}$ is the inflow of plasma from small arteries to capillaries and $Q_{\mathrm{pl,out,sc}}$ is the outflow of plasma from capillaries to small veins.

Finally, the mass of solute s in the RBC compartment in the capillaries changes at the rate described as:

$$\frac{dM_{s,rc,sc}}{dt} = -Q_{s,rc,sc} + c_{s,rc,sa}Q_{rc,in,sc} - c_{s,rc,sc}Q_{rc,out,sc} \tag{2.47}$$

where $c_{s,rc,sa}$ is the concentration of solute s in the RBCs suspended in small arteries plasma, $Q_{rc,in,sc}$ is the inflow of RBCs from small arteries to capillaries and $Q_{rc,out,sc}$ is the outflow of RBCs from capillaries to small veins.

2.3.3 Capillary Filtration

According to the Starling hypothesis [106], fluid transfer across the capillary wall depends on the balance of the hydrostatic and osmotic pressures between the capillary lumen and the surrounding interstitium (Starling forces). The transcapillary fluid exchange was traditionally understood as a balance between filtration at the arterial side of capillaries and reabsorption at the venous end of capillaries or postcapillary venules [107]. However, according to the relatively new physiological concept, in the vast majority of tissues, under typical body conditions, there is a net filtration of fluid out of capillaries to the interstitium, with the vascular refilling taking place through the lymphatic system [108, 109].

For calculating fluid filtration (J_{cap}) from the capillary compartment to the interstitial compartment, the following equation was used [27, 94]:

$$J_{cap} = Lp\left[(P_{sc,mean} - P_{is}) - \sum_{p} \sigma_{p,cap}(\pi_{p,pl,sc} - \pi_{p,is}) - \sum_{s} \sigma_{s,cap}\varphi_s\left(\alpha_{s,cap}\frac{c_{s,pl,sc}}{F_{pl,sc}} - \frac{c_{s,is}}{F_{is}}\right)RT\right] \tag{2.48}$$

where Lp is the whole-body hydraulic conductivity (or permeability) of capillary walls (assumed constant), $P_{sc,mean}$ is the mean hydraulic pressure of systemic capillary blood, P_{is} is the hydrostatic pressure of the interstitial fluid, $\pi_{p,pl,sc}$ is the capillary plasma oncotic pressure (colloid osmotic pressure) of protein p (albumin or globulins), $\pi_{p,is}$ is the oncotic pressure of protein p in the interstitial fluid, φ_s is the osmotic activity coefficient of solute s, $\alpha_{s,cap}$ is the Gibbs-Donnan coefficient for ion s and RT is a constant (=19.3 mmHg/mmol/L).

The above equation includes the impact of osmotic pressure of small solutes (the last term). Even though the concentrations of ions and small molecules on both sides of the capillary wall are normally equilibrated due to their high permeability and a very low reflection coefficient, transient transcapillary concentration gradients may occur during dialysis (especially for the solutes being removed in the dialyzer, such as urea or creatinine). For ions, the concentrations on both sides of the capillary wall are actually never equal due to the aforementioned Gibbs-Donnan effect. Since the additional osmotic pressure difference due to this effect is already included in the equations for the oncotic pressure (described below), the Gibbs-Donnan coefficient ($\alpha_{cap,s}$) was added in the above equation next to the plasma concentration of charged solutes in order not to account for this effect twice.

For simplicity, J_{cap} is treated in the model as pure water flow, i.e. the volume of non-reflected proteins and small solutes transported convectively with water is ignored as being negligible.

As far as the mean hydraulic pressure of capillary plasma ($P_{sc,mean}$) is concerned, based on experimental data found in [110–113], it was assumed that $P_{sc,mean}$ is resistant to isolated changes in arterial pressure (the autoregulatory capacity of the capillary bed), whereas 80% of changes in venous pressure are transmitted to the capillaries. $P_{sc,mean}$ is hence calculated as:

$$P_{sc,\,mean} = P_{sc,mean,0} + w_v \cdot (P_{sv} - P_{sv,0}) \tag{2.49}$$

where $P_{sc,mean,0}$ is the initial mean capillary pressure calculated from the initial steady-state conditions (see Sect. 2.5.1), w_v is a parameter (assumed 0.8) and $P_{sv,0}$ is the assumed initial (normal) pressure in the small veins compartment (see Table 2.6).

Note that with the above approach, $P_{sc,mean}$ is independent from isolated changes in arterial pressure (when venous pressure remains unchanged, as in the cited experiments), but in the real closed-loop circulatory system, it still depends indirectly on the arterial pressure, as a change in arterial pressure will likely affect the venous pressure.

The hydrostatic pressure of the interstitial fluid was described as a linear function of the interstitial volume [4, 27]:

$$P_{is} = P_{is,n} + \frac{1}{C_{is}} \cdot (V_{is} - V_{is,n}) \tag{2.50}$$

where $P_{is,n}$ is the normal interstitial pressure corresponding to the normal interstitial volume ($V_{is,n}$) and C_{is} is the interstitial compliance, which was assumed to be 12% of normal interstitial volume per mm Hg [114] (see Table 2.4).

As described in Sect. 2.2.1, the changes of interstitial pressure affect the transmural blood pressure in all extrathoracic vascular compartments and hence affect the volumes of these compartments.

To calculate the oncotic pressures exerted by albumin and globulins in the capillary plasma and the interstitial fluid, the following approach was used.

Landis and Pappenheimer [12] proposed the following equations to estimate the colloid osmotic pressure (or oncotic pressure) of human albumin solution and whole plasma for protein concentration in the range 0–25 g/dL (T = 37 °C, pH = 7.4, electrolytes, 0.15 M NaCl):

$$\pi_{\text{alb}} = 2.8\,c + 0.18\,c^2 + 0.012\,c^3 \tag{2.51}$$

$$\pi_{\text{pl}} = 2.1\,c + 0.16\,c^2 + 0.009\,c^3 \tag{2.52}$$

where π is expressed in mm Hg and c is the protein concentration in g/dL (albumin or total protein concentration, respectively).

The first term in the above equations corresponds to the van't Hoff's equation for the osmotic pressure of ideal solutions and gases [15, 117]:

$$\pi = cRT \tag{2.53}$$

where c is the concentration expressed in mmol/L, R is the ideal gas constant and T is absolute temperature ($RT = 19.3$ mm Hg/mmol/L at 37 °C).

The second and third terms in the above equations represent the Gibbs-Donnan effects (i.e. the interactions between the proteins and ions) and protein-protein interactions [12]. These two effects, especially the latter, are responsible for the rapid deviation of plasma osmotic pressure from the van't Hoff's equation for ideal solutes at higher protein concentrations (see Fig. 2.8).

Assuming that globulin-albumin interactions in plasma do not differ from albumin-albumin interactions, the oncotic pressure of plasma (π_{pl}) can be expressed as a weighted sum of the oncotic pressures exerted by albumin (π_{alb}) and globulins (π_{glob}) [12, 117]:

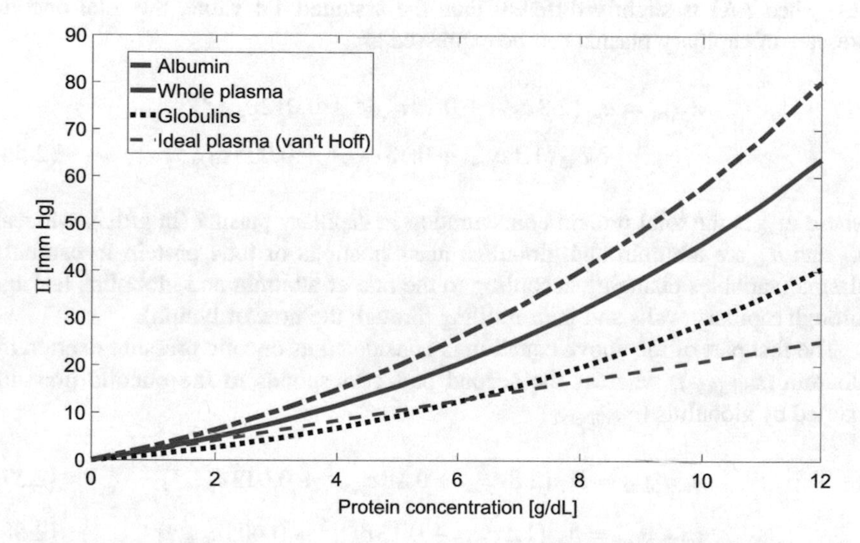

Fig. 2.8 Oncotic pressure-protein concentration curves for whole plasma (solid blue line), albumin (dash-dotted red line) and globulins (dotted black line) according to Eqs. (2.52, 2.51 and 2.55) for $T = 37$ °C, pH $= 7.4$, electrolytes: 0.15 M NaCl. The dashed blue line corresponds to the ideal limiting law of van't Hoff (Eq. 2.53) for whole plasma

$$\pi_{\mathrm{pl}} = a \cdot \pi_{\mathrm{alb}} + b \cdot \pi_{\mathrm{glob}} \tag{2.54}$$

where a and b are albumin and globulins mass fractions of total protein ($a + b = 1$).

Based on Landis-Pappenheimer equations for π_{alb} and π_{pl} (Eqs. 2.51 and 2.52), Nitta et al. [117] proposed two equations for π_{glob} satisfying the above relationship in the case when the albumin/globulin mass ratio (A/G) is 1.1 (as suggested originally by Landis and Pappenheimer) or 1.8 (as found by Nitta et al. in 13 healthy Japanese adults [117]).

Based on the approach by Nitta et al. [117], the following alternative equation for π_{glob} was derived assuming A/G = 1.4, which is a value between those used by Landis and Pappenheimer or Nitta et al. as used in this study (see Table 2.8), as seen in dialysis patients [123]:

$$\pi_{\mathrm{glob}} = 1.1\,c + 0.13\,c^2 + 0.005\,c^3 \tag{2.55}$$

The above equation for π_{glob} satisfies the Eq. (2.54) for A/G = 1.4 ($a = 0.58$ and $b = 0.42$) and π_{alb} and π_{pl} defined by Eqs. (2.51) and (2.52).

As indicated by Landis and Pappenheimer, given the large heterogeneity of globulins (molecular weight ranging from 45 to 1000 kDa [12]) and their different osmotic activities, the above equation should not be treated as representative of any particular fraction of globulins, but reflects the average osmotic properties of all non-albumin proteins.

Assuming that the original Landis-Pappenheimer equation for π_{alb} (Eq. 2.51) and the above equation for π_{glob} can be used to estimate the total plasma oncotic pressure also when A/G is slightly different than the assumed 1.4 value, the total oncotic pressure of capillary plasma can be expressed as:

$$\begin{aligned} \pi_{\mathrm{pl,\,sc}} = {} & a_{\mathrm{sc}}\left(2.8\,c_{\mathrm{p,\,sc}} + 0.18\,{c_{\mathrm{p,\,sc}}}^2 + 0.012\,{c_{\mathrm{p,\,sc}}}^3\right) \\ & + b_{\mathrm{sc}}\left(1.1\,c_{\mathrm{p,\,sc}} + 0.13\,{c_{\mathrm{p,\,sc}}}^2 + 0.005\,{c_{\mathrm{p,\,sc}}}^3\right) \end{aligned} \tag{2.56}$$

where $c_{\mathrm{p,sc}}$ is the total protein concentration in capillary plasma (in g/dL), whereas a_{sc} and b_{sc} are albumin and globulins mass fractions of total protein in capillary plasma (variables changing according to the rate of albumin and globulins leakage through capillary walls and their refilling through the flow of lymph).

The first part of the above equation is considered as oncotic pressure exerted by albumin ($\pi_{\mathrm{alb,pl,sc}}$), whereas the second part corresponds to the oncotic pressure exerted by globulins ($\pi_{\mathrm{glob,pl,sc}}$).

$$\pi_{\mathrm{alb,\,pl,\,sc}} = a_{\mathrm{sc}}\left(2.8\,c_{\mathrm{p,\,sc}} + 0.18\,{c_{\mathrm{p,\,sc}}}^2 + 0.012\,{c_{\mathrm{p,\,sc}}}^3\right) \tag{2.57}$$

$$\pi_{\mathrm{glob,\,pl,\,sc}} = b_{\mathrm{sc}}\left(1.1\,c_{\mathrm{p,\,sc}} + 0.13\,{c_{\mathrm{p,\,sc}}}^2 + 0.005\,{c_{\mathrm{p,\,sc}}}^3\right) \tag{2.58}$$

Similar equations were used for calculating the oncotic pressure of interstitial albumin and globulins based on total concentration of proteins in the interstitial compartment ($c_{p,is}$ in g/dL calculated according to Eq. 2.61) and the corresponding a_{is} and b_{is} fractions (also variable).

$$\pi_{is} = a_{is}\left(2.8\,c_{p,is} + 0.18\,{c_{p,is}}^2 + 0.012\,{c_{p,is}}^3\right) + b_{is}\left(1.1\,c_{p,is} + 0.13\,{c_{p,is}}^2 + 0.005\,{c_{p,is}}^3\right) \quad (2.59)$$

Note that the interstitial protein concentration at the same time both determines the capillary filtration (through $\pi_{p,is}$ in the Eq. 2.48) and depends on it (the higher the filtration rate, the more diluted the interstitial fluid and the lower the interstitial protein concentration).

The above approach for calculating the interstitial oncotic pressure ignores the most recent theory, according to which a protein gradient develops between the bulk interstitial fluid and the fluid directly beneath the glycocalyx (the semipermeable layer of endothelium, approximately 1 μm thick) [109, 124]. This is due to the high reflection coefficient for proteins (meaning that the ultrafiltrate passing through the capillary walls has a very low protein concentration and hence it has a diluting effect) as well as due to the relatively high fluid velocity through the intercellular cleft of continuous (non-fenestrated) capillaries resulting in a hindered protein diffusion from the bulk interstitial fluid (the latter effect being less significant in the fenestrated capillaries) [109, 124]. According to this theory, to calculate Starling forces across the capillary wall, instead of using the oncotic pressure (π_{is}) of the bulk interstitial fluid, one should use the (lower) oncotic pressure of the subglycocalyx fluid [108, 109]. Since the interstitial protein gradient develops especially at high filtration rates, while being relatively modest during normal conditions [124], for simplicity it was ignored, and the classic Starling approach was used instead (especially that during HD, due to reduced blood volume and increased plasma oncotic pressure, the filtration rate will be usually smaller than normal, at times becoming negative, when the fluid is transiently absorbed from the interstitium).

2.3.4 *Protein Transport*

Unlike in some previous models, in which it was assumed that the capillary wall is impermeable to plasma proteins [4], the proposed model includes the transcapillary leakage of proteins from plasma to interstitium as well as the convective protein refilling via the lymphatic system. The analysed proteins include the main plasma proteins: (1) albumin and (2) all globulins treated collectively (previous models typically analyse all plasma proteins represented by a single albumin-size molecule [27, 125, 126]). For simplicity, fibrinogen was treated as part of globulins, given its low concentration compared to albumin or globulins [127], and hence globulins represent all non-albumin plasma proteins.

The mass concentration of protein p (albumin or globulins) in the capillary plasma is calculated simply as the ratio of protein mass and capillary plasma volume:

$$c_{\mathrm{p,pl,sc}} = \frac{M_{\mathrm{p,pl,sc}}}{V_{\mathrm{pl,sc}}} \tag{2.60}$$

The molar concentrations of proteins are calculated assuming the molecular weight of albumin of 69 kDa and the average molecular weight of globulins of 170 kDa [127]. These assumed values of protein molecular weights agree well with the values estimated by equating the first term from the equations for plasma oncotic pressure exerted by albumin and globulins presented in the previous chapter (Eqs. 2.51 and 2.55) with the osmotic pressure of ideal protein solutions governed by the van't Hoff's law (Eq. 2.53) [117] (the thusly estimated molecular weights of albumin and globulins are approximately 69 kDa and 175 kDa, respectively).

Within the interstitial volume, a gel subvolume was identified, an excluded interstitial volume ($V_{\mathrm{is,ex}}$), which, due to the interactions of the polysaccharide molecules, is not available for the macromolecules such as proteins, thus increasing the effective protein concentration in the interstitial fluid [26, 128]. It was assumed in the model that the volume of this excluded space (50% of normal interstitial volume [114]) remains constant at all times, thus ignoring its possible reduction in case of tissue overhydration [128]. Therefore, the concentration of proteins in the interstitial fluid is given by the following equation:

$$c_{\mathrm{p,is}} = \frac{M_{p,\mathrm{is}}}{V_{\mathrm{is}} - V_{\mathrm{is,ex}}} \tag{2.61}$$

For modelling the transport of proteins across the capillary wall the standard diffusive-convective flow equation was used [99, 100]:

$$Q_{\mathrm{p}} = \mathrm{PS}_{\mathrm{p}}\left(c_{\mathrm{p,pl,sc}} - c_{\mathrm{p,is}}\right) + J_{\mathrm{cap}}\left(1 - \sigma_{\mathrm{p,cap}}\right)\left[(1-f)c_{\mathrm{p,pl,sc}} + fc_{\mathrm{p,is}}\right] \tag{2.62}$$

where p denotes albumin or globulins, J_{cap} is the rate of fluid filtration from the capillaries to the interstitium, $\sigma_{p.\mathrm{cap}}$ is the capillary reflection coefficient of protein p, PS_{p} is the permeability-surface product of the lumped capillary wall for protein p (assumed constant) and f is defined as in the Eq. (2.35).

The above assumes a homogenous membrane of the capillary wall available for protein transport and hence does not take into account the differences between the small and large capillary pores, as used in some other models [125, 129].

Equation (2.62) is also used for the case of possible transient absorption of fluid from the interstitium to the capillaries (when $J_{\mathrm{cap}} < 0$).

Protein transport between the interstitium and tissue cells as well as between plasma and RBCs was neglected. Similarly, the model neglects the possible protein leakage at the dialyzer membrane. It was also assumed that during the dialysis session, the protein catabolic rate is equal to the rate of protein synthesis, and hence the total amount of proteins in the body remains constant.

The difference in protein concentration across the capillary wall results in the Gibbs-Donnan effect included in the solute kinetics equations (c.f. Sect. 2.3.2).

2.3.5 Lymph Flow

The absorption of lymph from the interstitium (afferent lymph) was described by linear functions of the interstitial pressure, as done by Gyenge et al. [27]:

$$Q_{\mathrm{L}} = Q_{\mathrm{L,n}} + \mathrm{LS} \cdot (P_{\mathrm{is}} - P_{\mathrm{is,n}}), \quad P_{\mathrm{is}} \geq P_{\mathrm{is,n}} \tag{2.63}$$

$$Q_{\mathrm{L}} = Q_{\mathrm{L,n}} \frac{(P_{\mathrm{is}} - P_{\mathrm{is,ex}})}{(P_{\mathrm{is,n}} - P_{\mathrm{is,ex}})}, \quad P_{\mathrm{is,n}} \geq P_{\mathrm{is}} \geq P_{\mathrm{is,ex}} \tag{2.64}$$

where $Q_{\mathrm{L,n}}$ is the normal steady-state lymph flow corresponding to the normal interstitial fluid pressure $P_{\mathrm{is,n}}$, LS is the lymph flow sensitivity to interstitial pressure changes and $P_{\mathrm{is,ex}}$ is the pressure of the interstitial fluid when its volume decreases to the volume excluded to proteins ($V_{\mathrm{is,ex}}$), at which point the lymph flow ceases.

In a normal human, up to half of the afferent lymph flow is reabsorbed by the venules at lymph nodes as protein-free fluid, whereas the other half, i.e. the postnodal efferent lymph (with an increased protein concentration compared to the interstitial fluid), is drained to the large veins [109, 115]. Since in the proposed model it does not make much difference whether the whole lymph is drained to the large veins or if part of the lymph is drained earlier to the small veins, it was assumed that all lymph is drained to the large veins, thus neglecting the possible reabsorption of the afferent lymph by venules at lymph nodes. Since the volume of the lymphatic system was neglected, it was assumed that the fluid transfer from the interstitium to the veins occurs instantaneously.

2.3.6 Cellular Water Transport

Water is exchanged between the intracellular and interstitial compartments according to the crystalloid and colloid osmotic pressure gradients (with both gradients balancing each other in the steady-state equilibrium). The water flow out of the cells can be hence expressed as [4, 19]:

$$J_{\mathrm{cell}} = K_{\mathrm{w,cell}}(O_{\mathrm{is}} - O_{\mathrm{ic}}) \tag{2.65}$$

where $K_{\mathrm{w,cell}}$ is the whole-body water transfer coefficient of the cellular membrane and O_{is} and O_{ic} are interstitial and intracellular effective osmolarities (in mOsm/L) given by the following formulae [19]:

$$O_{\mathrm{is}} = \sum_{\mathrm{s}} \sigma_{\mathrm{s,cell}} \cdot \varphi_{\mathrm{s}} \cdot \frac{c_{\mathrm{s,is}}}{F_{\mathrm{is}}} \tag{2.66}$$

$$O_{\mathrm{ic}} = \sum_{\mathrm{s}} \sigma_{\mathrm{s,cell}} \cdot \varphi_{\mathrm{s}} \cdot \frac{c_{\mathrm{s,ic}}}{F_{\mathrm{ic}}} \tag{2.67}$$

where $\sigma_{s,cell}$ is the reflection coefficient of solute s (reducing the effective osmotic pressure difference across the cellular membrane when $\sigma_{s,cell} < 1$), φ_s is the osmotic coefficient representing the solute interactions reducing the effective osmolarity [14] and F_{ic} and F_{is} are variable water fractions of the intracellular and interstitial fluid, respectively (used for all solutes except for proteins).

The interstitial and intracellular osmolarities include the osmotic impact of all solutes being considered in the model, including unspecified cations and anions, as well as proteins (other solutes not included in the model explicitly were ignored assuming their minor contribution to total osmolarity or minor changes of concentration during dialysis, or both). $\varphi = 0.93$ was assumed for ions and $\varphi = 1$ for uncharged solutes [121] (assuming that the osmotic coefficients are independent from solute concentrations). For the interstitial proteins, the osmotic coefficient was calculated based on the interstitial oncotic pressure calculated earlier (π_{is}) and the interstitial total protein concentration (in mmol/L) using the van't Hoff's law [15]:

$$\varphi_{\mathrm{p,is}} = \frac{1}{19.3} \frac{\pi_{\mathrm{is}}}{c_{\mathrm{p,is}}} \tag{2.68}$$

For the intracellular proteins, due to the lack of a similar equation for the intracellular oncotic pressure and a relatively stable intracellular protein concentration during HD, a constant osmotic coefficient $\varphi_{p,ic} = 2$ was assumed [130].

The reflection coefficient ($\sigma_{s,cell}$) was effectively used for urea and creatinine only, while for all other solutes, it was assumed equal to 1 (i.e. it was assumed that water channels are not used by electrolytes [131] and that proteins do not cross the cellular membrane).

It was assumed that there is no hydrostatic pressure difference across the cellular membrane [15].

Volume changes of interstitial fluid can be expressed as:

$$\frac{\mathrm{d}V_{\mathrm{is}}}{\mathrm{d}t} = J_{\mathrm{cell}} + J_{\mathrm{cap}} + Q_{\mathrm{p,diff,V}} - Q_{\mathrm{L}} \tag{2.69}$$

where J_{cell} is water flow from tissue cells to interstitium (Eq. 2.65), J_{cap} is fluid filtration from capillaries to interstitium (Eq. 2.48), Q_L is the lymph absorption from the interstitium (see Eqs. 2.63 and 2.64) and $Q_{p,diff,V}$ is the volumetric flow of protein diffusion from capillary plasma to interstitium calculated as:

$$Q_{\mathrm{p,diff,V}} = \mathrm{PS_p}\left(c_{\mathrm{p,pl,sc}} - c_{\mathrm{p,is}}\right) \cdot \frac{\mathrm{MW_p}}{\rho_{\mathrm{p}}} \tag{2.70}$$

where ρ_p is the protein density (assumed 1.37 g/cm^3 for both albumin and globulins [132]).

Within each blood compartment, there is also additional water exchange between plasma and RBCs. For the i-th blood compartment, the flow of water from plasma to RBCs can be expressed as follows (c.f. Eq. 2.65):

$$Q_{\mathrm{pl-rc},i} = -K_{\mathrm{w,rc}} S_{\mathrm{rc},i} \left(O_{\mathrm{pl},i} - O_{\mathrm{rc},i} \right) \tag{2.71}$$

where $K_{w,rc}$ is the water transfer coefficient of the RBC membrane per unit surface, $S_{rc,i}$ is the total surface area of RBCs in the i-th compartment and $O_{pl,i}$ and $O_{rc,i}$ are osmolarities of plasma and RBC fluid in the i-th compartment (in mOsm/L), calculated similarly as for the interstitial and intracellular fluids (c.f. Eqs. 2.66 and 2.67):

$$O_{\mathrm{pl},i} = \sum_{\mathrm{s}} \sigma_{\mathrm{s,rc}} \cdot \varphi_{\mathrm{s}} \cdot \frac{c_{\mathrm{s,pl},i}}{F_{\mathrm{pl},i}} \tag{2.72}$$

$$O_{\mathrm{rc},i} = \sum_{\mathrm{s}} \sigma_{\mathrm{s,rc}} \cdot \varphi_{\mathrm{s}} \cdot \frac{c_{\mathrm{s,rc},i}}{F_{\mathrm{rc},i}} \tag{2.73}$$

where $\sigma_{s,rc}$ is the reflection coefficient of solute s at the RBC membrane and $F_{pl,i}$ and $F_{rc,i}$ are plasma and RBC water fractions in the i-th blood compartment (used for all solutes except for proteins) changing from the initial levels ($F_{pl,0}$ and $F_{rc,0}$) according to plasma and RBC water content changes (tracked separately).

Similarly as for the interstitial proteins (Eq. 2.68), the osmotic coefficient of plasma proteins in the i-th compartment is based on plasma oncotic pressure ($\pi_{pl,i}$) calculated according to Eq. (2.52) and the total plasma protein concentration (in mmol/L) using the van't Hoff's law [15]:

$$\varphi_{\mathrm{p,pl},i} = \frac{1}{19.3} \frac{\pi_{\mathrm{pl},i}}{c_{\mathrm{p,pl},i}} \tag{2.74}$$

The osmotic coefficient of RBC proteins (assumed to be represented by haemoglobin) depends on RBC haemoglobin concentration (in mmol/L) as follows [121]:

$$\varphi_{\mathrm{Hb,rc},i} = 1 + 0.0645 \cdot c_{\mathrm{Hb,rc},i} + 0.0258 \cdot c_{\mathrm{Hb,rc},i}^2 \tag{2.75}$$

2.3.7 Dialyzer Mass Exchange

Within the dialyzer the exchange of ions and low-weight molecules takes place between plasma and dialysate fluid (typically with a counter-current flow arrangement). For all small solutes except other anions, the diffusive-convective flux of substance s across the dialyzer membrane can be expressed as follows [21]:

$$Q_{s,d} = D_s \Delta c + Q_{uf}(1 - \sigma_{s,d})\alpha_{s,d}\frac{c_{s,pl,out}}{F_{pl,d}} \tag{2.76}$$

where D_s is the diffusive dialysance (or clearance) of solute s (when there is no ultrafiltration), Δc is the effective concentration difference between plasma and dialysate fluid driving the diffusion process, Q_{uf} is the dialyzer ultrafiltration rate, $\sigma_{s,d}$ is the reflection coefficient of solute s at the dialyzer membrane (assumed 0 for all small solutes), $\alpha_{s,d}$ is the Gibbs-Donnan coefficient for ion s, $c_{s,pl,out}$ is the concentration of solute s in plasma leaving the dialyzer compartment and $F_{pl,d}$ is plasma water fraction in the dialyzer compartment (variable).

Δc was defined as the difference between the concentration of solute s in the plasma entering the dialyzer ($c_{s,pl,in}$), corrected for plasma water fraction ($F_{pl,in}$) and the Gibbs-Donnan effect, and the concentration of solute s in dialysate fluid inflow ($c_{s,d,in}$) [21]:

$$\Delta c = \alpha_{s,d}\frac{c_{s,pl,in}}{F_{pl,in}} - c_{s,d,in} \tag{2.77}$$

Since the Gibbs-Donnan coefficients at the dialyzer membrane are much more important in the model than the Gibbs-Donnan coefficients at the capillary walls (given that the coefficients at the dialyzer membrane affect the amounts of ions exchanged with the dialysate fluid, whereas the coefficients at the capillary wall influence only the internal disequilibrium of ion concentrations between plasma and interstitial fluid), instead of using the theoretical values for mono- or divalent cations or anions, the values of the Gibbs-Donnan coefficients for the dialyzer membrane were taken from the empirical literature studies on blood serum and a protein-free fluid on both sides of a semipermeable membrane. The following values were assumed based on data found in [101, 102]: 0.94 for sodium and potassium, 1.01 for chloride and bicarbonate and 0.9 for other (divalent) cations.

It was assumed that, as free (unbounded) solutes leave plasma to the dialysate, there is no unbinding of previously bounded solutes, and hence the free concentration of solute s in plasma obtained by the use of the Gibbs-Donnan coefficient ($\alpha_{s,d}$) slightly underestimates the actual concentration difference between plasma and dialysate fluid [21].

The solute transport across the dialyzer membrane is possible in both directions and depends on the magnitude of the ultrafiltration and the concentration difference between plasma inflow and dialysate fluid inflow.

Simultaneously, blood plasma flowing through the dialyzer exchanges solutes and water with the RBCs suspended in plasma (modelled separately as in all cardiovascular compartments), and hence a fraction of water and solutes entering the dialysate fluid comes effectively from the RBCs flowing through the dialyzer.

Similarly as for other membranes, the flux of other anions across the dialyzer membrane is provided to maintain electroneutrality.

2.3.8 Blood Mixing

Under steady-state conditions, the composition of blood is almost uniform across the whole CVS (except for the assumed haematocrit variation and small differences in solute concentrations due to capillary filtration and lymph drainage). Blood plasma is also in osmotic and diffusive equilibrium with the interstitial fluid and with the erythrocytes. During HD, however, two blood compartments participate in intense mass exchange processes – capillary compartment exchanging water and solutes with the interstitium and dialyzer compartment exchanging water and small solutes with dialysate. Blood flowing out of the dialyzer is haemoconcentrated (due to ultrafiltration) and depleted from some solutes (due to diffusive and convective loss of small solutes across the dialyzer membrane). This blood mixes then with venous blood in large veins (for simplicity, mixing with access blood flow was ignored), passes through the entire cardiovascular system and enters systemic capillaries, where it can absorb more solutes from the interstitium (except for blood bypassing microcirculation through the vascular access). Mixing of blood and equilibration of solute concentrations occurs, hence, throughout the whole CVS as well as in the extracorporeal circuit, with all solute transport between the compartments occurring convectively (the diffusion was neglected). Assuming homogeneity of all blood compartments, in dynamic states the concentrations of solutes change stepwise between the compartments, whereas in the steady-state equilibrium, the concentrations are almost equal in all plasma and all RBC compartments.

It was assumed that mixing of solutes within each compartment is obtained instantaneously, and hence there is no concentration gradient within the compartments. In particular, it was assumed that there are no gradients of plasma solute concentrations along the represented compartment, and that, at any given time, all RBCs within the given blood compartment have identical (average) properties. This simplification (associated with lumped parameter modelling) should not cause large inaccuracies, given the relatively large number of vascular compartments considered in the model and the duration of HD.

As indicated in Sect. 2.2.5, the model distinguishes between the flow of plasma and the flow of RBCs. For each compartment there is a separate mass of solute s in the plasma and in RBCs. Based on dynamic changes of the total volume of RBCs within each blood compartment as well as changes of the mass of each solute contained in the given lumped RBC compartment, the instantaneous mass and molar concentration of the given solute in the RBCs are calculated for each blood compartment.

The mass of solute s in the lumped RBC compartment within the k-th blood compartment changes as a result of the inflow of solute within the RBC inflow from the preceding ($k-1$-th) compartment, the outflow of solute within the RBC outflow to the subsequent ($k+1$-th) compartment and the solute exchange across the RBC membrane (calculated according to Eq. 2.41):

$$\frac{\mathrm{d}M_{\mathrm{s,rc},k}}{\mathrm{d}t} = c_{\mathrm{s,rc},k-1}Q_{\mathrm{rc,in},k} - c_{\mathrm{s,rc,k}}Q_{\mathrm{rc,out},k} - Q_{\mathrm{s,rc},k} \tag{2.78}$$

where $Q_{\mathrm{rc,in},k}$ and $Q_{\mathrm{rc,out},k}$ are volumetric RBC inflow and outflow to and from the k-th compartment.

Similarly, for the changes of mass of solute s in the plasma compartment within the k-th blood compartment, we have:

$$\frac{\mathrm{d}M_{\mathrm{s,pl},k}}{\mathrm{d}t} = c_{\mathrm{s,pl,k}-1}Q_{\mathrm{pl,in},k} - c_{\mathrm{s,pl,k}}Q_{\mathrm{pl,out},k} + Q_{\mathrm{s,rc},k} \tag{2.79}$$

The total volume of the RBC compartment within the k-th blood compartment changes according to the RBC volumetric inflow ($Q_{\mathrm{rc,in},k}$) and outflow ($Q_{\mathrm{rc,out},k}$) to and from the k-th compartment as well as due to water exchange between blood plasma and RBCs within the given compartment ($Q_{\mathrm{pl-rc},k}$, see Eq. 2.71):

$$\frac{\mathrm{d}V_{\mathrm{rc},k}}{\mathrm{d}t} = Q_{\mathrm{rc,in},k} - Q_{\mathrm{rc,out},k} + Q_{\mathrm{pl-rc},k} \tag{2.80}$$

Similarly, for the changes of plasma volume within the k-th compartment, we have:

$$\frac{\mathrm{d}V_{\mathrm{pl},k}}{\mathrm{d}t} = Q_{\mathrm{pl,in},k} - Q_{\mathrm{pl,out},k} - Q_{\mathrm{pl-rc},k} \tag{2.81}$$

For some compartments the above equations include additionally other inflows or outflows, as shown in Fig. 2.7. For the capillary plasma, there is an additional outflow/inflow corresponding to fluid filtration/absorption to/from the interstitium (J_{cap}). For the dialyzer plasma, there is an outflow corresponding to ultrafiltration (Q_{uf}). For the large arteries and large veins, there is an additional outflow and inflow corresponding to blood flow through the arteriovenous access. For the large veins, there is an additional fluid inflow from the lymph vessels.

An equation similar to Eq. (2.79) was used to describe the changes of mass of protein p in each plasma compartment, except that it was assumed that proteins do not cross the RBC membrane:

$$\frac{\mathrm{d}M_{\mathrm{p,pl},k}}{\mathrm{d}t} = c_{\mathrm{p,pl},k-1}Q_{\mathrm{pl,in},k} - c_{\mathrm{p,pl},k}Q_{\mathrm{pl,out},k} \tag{2.82}$$

For the capillary compartment, there is an additional component to the above equation corresponding to protein leakage (Q_{p} defined by Eq. 2.62), while for large veins there is a component describing protein refilling via the lymphatic system.

2.3.9 Model Integration

The integration of the models of cardiovascular system, baroreflex and whole-body water and solute kinetics creates a relatively complex network of direct or indirect relationships between a large number of model variables. Figure 2.9 shows the main of such relationships between the cardiovascular variables (cardiac performance parameters, vascular pressures, volumes and resistances) and the variables describing the extravascular fluid compartments. On top of these relationships, there is an equally complex dependency network between the volumes and osmolarities of individual fluid compartments governed by the inter-compartmental transport of solutes and water (for clarity not shown in Fig. 2.9, except for the transcapillary water filtration and the flow of lymph).

The presented model was designed to reflect relatively many fluid compartments and phenomena in an attempt to provide a reasonably accurate physiologically based model of a virtual patient (although still based on several assumptions and simplifications, as discussed earlier). The model was developed mainly for studying cardiovascular dynamics during HD, but it provides a comprehensive platform enabling detailed analyses of different aspects of dialysis treatment, such as kinetics of individual solutes, changes of intracellular and extracellular volume or the impact of fistula blood flow or cardiopulmonary recirculation on dialysis efficiency. The presented model can be also employed for analysing transport processes and phenomena occurring in the human body in response to various non dialysis-related perturbations, such as fluid infusions, or simply for a quantitative model-based analysis of certain aspects of human physiology. Of course for certain specific modelling tasks (such as modelling kinetics of a given substance, e.g. urea), the presented model can be simplified in many ways, such as combining several vascular compartments into one compartment, modelling a limited number of substances or neglecting some of the modelled features (e.g. the dependence between the vessel volume and its resistance to blood flow or variation of haematocrit across the cardiovascular system). However, for a thorough analysis of processes affecting blood volume changes (and hence blood pressure), a full integration of haemodynamic model with the model of solute and water kinetics seems indispensable. Moreover, thanks to the description of the whole-body distribution and transport of water and individual solutes, in particular the distinction between the intracellular and extracellular fluids, the model could be used in the future for analyses of different dialysis modalities with variable dialysis parameters (e.g. profiling of sodium in the dialysate fluid or profiling of the ultrafiltration rate) in order to enhance the removal of water from the cells during dialysis, while keeping the proper balance of sodium, potassium and other electrolytes, as well as maintaining cardiovascular stability.

The proposed level of accuracy of some of the modelled features may not be needed for modelling a standard dialysis session with limited and relatively smooth changes of analysed variables, but it may turn out important for analysing more intense dialysis sessions with large dynamic haemodynamic or fluid changes; for

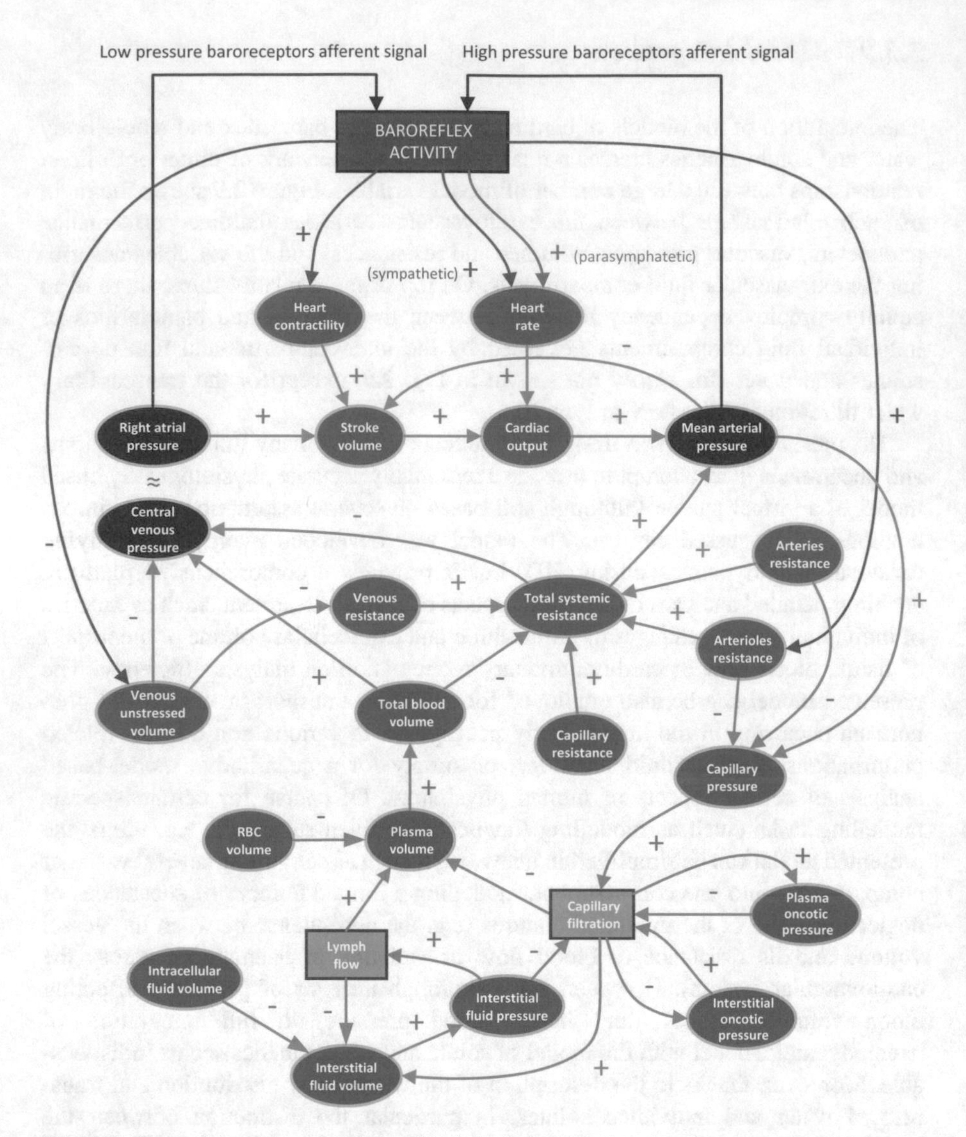

Fig. 2.9 The main positive (+) and negative (−) relationships between various cardiovascular and extravascular variables. Red ovals represent the input signals to baroreflex mechanisms; green rectangles represent non-state variables describing the flow of matter (lymph flow or capillary fluid filtration)

instance, a nonlinear relationship between atrial pressure and stroke volume (the Frank-Starling mechanism) could be replaced by a simple linear relationship for cases with low changes in blood volume and hence low changes in atrial pressure. On the other hand, some cases may require a more accurate representation of transport processes in order to better reflect the clinically observed phenomena. Therefore, the presented model should be treated as a framework for various modelling efforts, some of which may require model improvements or extensions, while for other the model could be simplified (see Chap. 5 for the discussion on some possible directions for future modelling studies).

2.4 Parameter Assignment

2.4.1 Cardiovascular System Parameters

All parameters of the CVS model were assigned on the basis of data found in the physiological literature, similarly as done for the previous VM model [2], with some changes in the vascular compliances (see Table 2.1) due to the modified definition of arterial and venous compartments.

2.4.1.1 Cardiovascular Compliances

The compliances of the cardiac compartments were taken as the sum of diastolic compliances of the corresponding atrium and ventricle as follows [46]: 0.25 ml/mmHg/kg for the right heart and 0.14 ml/mmHg/kg for the left heart. The compliances of the vascular compartments were chosen based on the following data [133]: (1) the total human vascular compliance is 2.08 ml/mmHg/kg, (2) the total pulmonary compliance is 0.45 ml/mmHg/kg and (3) the compliance of the systemic arteries is 0.034 ml/mmHg/kg. The distinction between the compliance of the pulmonary arteries and pulmonary veins was based on data from Tanaka et al. [48] taking the pulmonary arterial compliance equal to 0.14 ml/mmHg/kg. Since the compliance of the pulmonary veins in the experiments of Tanaka et al. included the compliance of

Table 2.1 Compliances of the cardiovascular compartments [mL/mmHg/kg] (some values correspond to two or more compartments combined)

<table>
<tr><th>CVS compartment</th><th>[5]</th><th>[45]</th><th>[46]</th><th>[47]</th><th>[48]</th><th>This study</th></tr>
<tr><td>Large arteries (incl. aorta)</td><td rowspan="2">0.057</td><td rowspan="2">0.034</td><td>–</td><td>0.0228</td><td>–</td><td>0.0323</td></tr>
<tr><td>Small arteries (incl. arterioles)</td><td>–</td><td>0.0012</td><td>–</td><td>0.0017</td></tr>
<tr><td>Systemic capillaries</td><td>–</td><td rowspan="4">1.596</td><td>–</td><td>–</td><td>–</td><td>0.0030</td></tr>
<tr><td>Small veins (incl. venules)</td><td rowspan="2">1.588</td><td>–</td><td>–</td><td>–</td><td>0.7715</td></tr>
<tr><td>Large veins (incl. vena cavae)</td><td>–</td><td>–</td><td>–</td><td>0.7715</td></tr>
<tr><td>Right atrium</td><td rowspan="2">0.45</td><td>0.05</td><td>–</td><td>–</td><td rowspan="2">0.25</td></tr>
<tr><td>Right ventricle</td><td>–</td><td>0.20</td><td>–</td><td>–</td></tr>
<tr><td>Pulmonary arteries</td><td>0.09</td><td rowspan="3">0.45</td><td>–</td><td>–</td><td>0.11</td><td>0.14</td></tr>
<tr><td>Pulmonary veins</td><td>0.36</td><td>–</td><td>–</td><td rowspan="2">0.26</td><td>0.28</td></tr>
<tr><td>Left atrium</td><td rowspan="2">0.27</td><td>0.03</td><td>–</td><td rowspan="2">0.14</td></tr>
<tr><td>Left ventricle</td><td>–</td><td>0.11</td><td>–</td><td>–</td></tr>
<tr><td>Total</td><td>2.82</td><td>2.08</td><td>–</td><td>–</td><td>–</td><td>2.39</td></tr>
</table>

the left atrium, the pulmonary venous compliance was set in this study to 0.28 ml/mmHg/kg (being the difference between the assumed total pulmonary compliance and the sum of the pulmonary arterial compliance and left atrial compliance). The compliance of the systemic capillaries was taken as 0.003 ml/mmHg/kg [59]. The distinction between the compliance of large and small arteries was made based on data from Wong et al. [47]. Total venous compliance was divided equally among large and small veins based on the results obtained by Spaan [134].

2.4.1.2 Cardiac Effectiveness

Parameters x and s of the stroke volume functions (separate for the right and left heart) were calculated so that under initial steady-state conditions, the ventricular outputs are equal to the assumed normal cardiac output (see Sect. 2.5.1) and the local slope of the SV curve is equal to the stroke volume sensitivity to atrial pressure. For the left heart, the latter was taken as 6 mL/mmHg based on data from the literature [135, 136]. For the right heart, the value 12 mL/mmHg was used assuming that the right heart is twice as sensitive to atrial pressure changes as the left heart [137] (Fig. 2.10, Table 2.2).

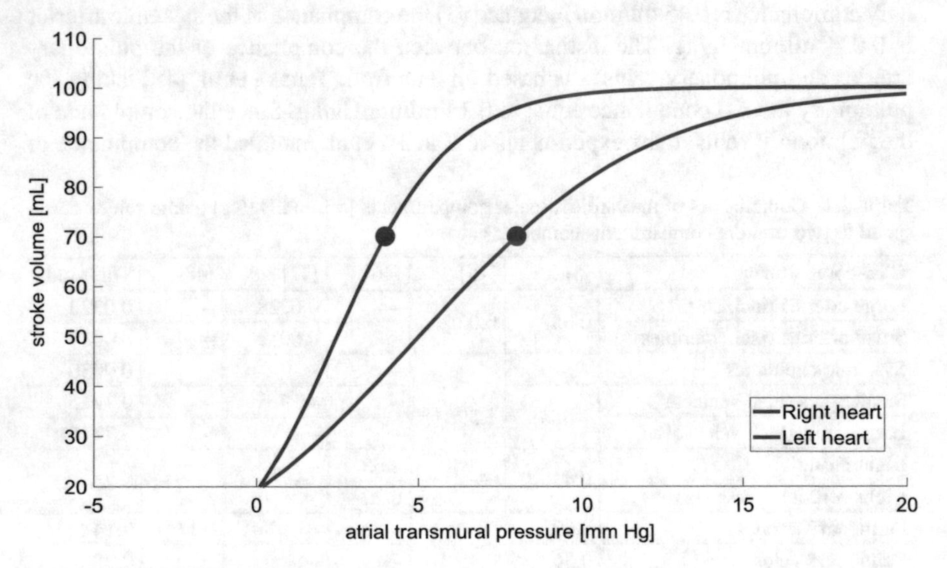

Fig. 2.10 Stroke volume-atrial pressure curves for the right and left heart assuming normal afterload

Table 2.2 Parameters describing the stroke volume-atrial pressure relationship for the left and right heart ventricles

Parameter	Symbol	Value	Unit	References
Maximal stroke volume	SV_{max}	100	mL	[138]
SV_r pressure offset	x_r	−2.52	mmHg	Calculated
SV_l pressure offset	x_l	−5.03	mmHg	Calculated
Central slope of the SV_r curve	s_r	1.75	–	Calculated
Central slope of the SV_l curve	s_l	3.50	–	Calculated
Afterload parameter	m	0.2	–	[2]

2.4.1.3 Baroreflex

The amplitudes and arterial gains of all baroreflex mechanisms (see Table 2.3) were assigned to reproduce the results of open loop experiments in vagotomized dogs [139–144], in which changes in arterial pressure, cardiac output, systemic resistance and venous capacity were observed in response to changes of pressure applied to isolated carotid sinuses. The amplitude and arterial gains of heart rate control were assigned to reproduce the neck chamber experiments in young humans [141]. The same experiments were used by Magosso et al. [77] and Ursino et al. [4] for assigning their baroreflex amplitudes and gains (due to previously discussed differences between the proposed model vs cited models, the values of parameters are slightly different). The cardiopulmonary gains of all mechanisms were taken as averages of gains used in the previous models [4, 77] (note the 0 cardiopulmonary gain for the heart rate control reflecting the negligible impact of cardiopulmonary baroreceptors in the control of heart rate compared to arterial baroreceptors).

The minimal and maximal values of all four sigmoidal functions were taken symmetrically around the normal (basal) value of the parameter being controlled. For the control of resistance of arterioles, the reference resistance was set to zero, and hence, depending on the circumstances, the baroreflex will increase or decrease the resistance of arterioles above or below the zero reference level, thus increasing or reducing the resistance of the compartment representing small arteries and arterioles. For the control of heart contractility, the reference value of contractility (parameter E from Eq. 2.11) was set to 1.

The time constants (τ) were taken as a compromise between the values from [5], [77] and [145]. The possible time delays were ignored, as in the previous VM model [2].

As done in [4], the values of parameters k describing the slope at the central point of the sigmoidal curves were calculated as ¼ of the respective response amplitude (dimensionless), which corresponds to the slope equal to 1 or −1 (depending on the direction of the sigmoidal function).

Table 2.3 Parameters of the baroreflex regulatory mechanisms based on signals from cardiopulmonary (c) and arterial (a) baroreceptors, controlling four system parameters: peripheral resistance (*R*), venous unstressed volume (*V*), heart contractility (*E*) and heart period (*T*), with the latter being controlled through both vagal (v) and sympathetic (s) mechanisms

Description	Parameter	Value	Unit	References
Amplitudes	ΔT	0.7	s	Assigned to match experimental data [139–144]
	ΔR	1.0	mmHg*s/mL	
	ΔV	600	mL	
	ΔE	0.5	–	
Time constants	τ_{Tv}	0.5	s	[5, 77, 145]
	τ_{Ts}	1.5	s	
	τ_{R}	3	s	
	τ_{V}	10	s	
	τ_{E}	6	s	
Slope determinants	k_{T}	0.175	s	Calculated (as in [4])
	k_{R}	0.25	mmHg*s/mL	
	k_{V}	150	mL	
	k_{E}	0.125	–	
Gains	$G_{c,Tv}$	0	s/mmHg	Cardiopulmonary gains: [4, 77] Arterial gains: assigned to match experimental data [139–144]
	$G_{a,Tv}$	0.02	s/mmHg	
	$G_{c,Ts}$	0	s/mmHg	
	$G_{a,Ts}$	0.01	s/mmHg	
	$G_{c,R}$	0.63	s/mL	
	$G_{a,R}$	0.03	s/mL	
	$G_{c,V}$	208	mL/mmHg	
	$G_{a,V}$	14	mL/mmHg	
	$G_{c,E}$	0.10	1/mmHg	
	$G_{a,E}$	0.01	1/mmHg	

2.4.2 Transport Parameters

The values of all parameters related to water and solute transport as well as the parameters of the interstitial fluid and lymph flow were taken from the literature (see Tables 2.4 and 2.5) with a few exceptions, as described below.

The parameters β describing the ratios between intracellular and extracellular concentrations of electrolytes in equilibrium were calculated from normal Na^+, K^+, Cl^- and HCO_3^- concentrations in the intracellular water (13, 140, 3 and 10 mmol/L, respectively) and in the interstitial fluid (145.3, 4.7, 114.7 and 26.5 mmol/L, respectively, corrected for the assumed interstitial water fraction of 0.98) [146]. A similar approach was used to calculate the parameters β for the RBC membrane,

Table 2.4 Parameters describing the interstitial fluid and the lymph absorption

Parameter	Symbol	Value	Unit	Reference
Normal interstitial pressure	$P_{is,n}$	−1.0	mmHg	[114]
Normal interstitial volume	$V_{is,n}$	14.97	L	See Sect. 2.5.1
Interstitial compliance	C_{is}	1.8	L/mmHg	Calculated [114]
Normal lymph absorption	$Q_{L,n}$	8.0	L/day	[93, 115]
Lymph flow sensitivity	LS	43.1	mL/mmHg/h	[116]

Table 2.5 Parameters of water and solute transport (the whole-body values)

	Parameter	Symbol	Value	Unit	References
Tissue cell membrane	Na^+ mass transfer	$K_{Na,cell}$	0.15	L/min	[18, 19]
	K^+ mass transfer	$K_{K,cell}$	0.01	L/min	[18, 19]
	Cl^- mass transfer	$K_{Cl,cell}$	0.20	L/min	Assumed
	HCO_3^- mass transfer	$K_{HCO3,cell}$	0.20	L/min	[18, 19]
	Urea mass transfer	$K_{U,cell}$	0.80	L/min	[20, 21]
	Creatinine mass transfer	$K_{Cr,cell}$	0.60	L/min	[20]
	Na^+ equilibrium ratio	$\beta_{Na,cell}$	0.088	–	Calculated
	K^+ equilibrium ratio	$\beta_{K,cell}$	29.19	–	Calculated
	Cl^- equilibrium ratio	$\beta_{Cl,cell}$	0.026	–	Calculated
	HCO_3^- equilibrium ratio	$\beta_{HCO3,cell}$	0.37	–	Calculated
	Reflection coefficient for ions	$\sigma_{ions,cell}$	1.0	–	Assumed
	Urea reflection coefficient	$\sigma_{U,cell}$	0.7	–	[22] (=RBC)
	Creatinine reflection coefficient	$\sigma_{Cr,cell}$	0.7	–	(=urea)
	Water transfer coefficient	$K_{w,cell}$	0.25	L^2/min/mmol	[18, 23]
Capillary wall	Urea permeability	p_U	2.2×10^{-5}	cm/s	[24, 25]
	Creatinine permeability	p_{Cr}	1.5×10^{-5}	cm/s	[24, 25]
	Ion permeability	p_{ions}	2.2×10^{-5}	cm/s	(=urea)
	Albumin permeability-surface product	PS_{alb}	246	mL/h	Calculated
	Globulins permeability-surface product	PS_{glob}	116	mL/h	Calculated
	Reflection coefficient for albumin	$\sigma_{alb,cap}$	0.95	–	[26, 27]
	Reflection coefficient for globulins	$\sigma_{glob,cap}$	0.98	–	Assumed
	Reflection coefficient for small solutes	$\sigma_{s,cap}$	0.05	–	[28]
	Total capillary surface area	S_{cap}	600	m^2	[14]
	Hydraulic conductivity	Lp	270	mL/h/mmHg	[29, 30]

(continued)

Table 2.5 (continued)

	Parameter	Symbol	Value	Unit	References
Red blood cell membrane	Na^+ permeability	$p_{Na,rc}$	4.0×10^{-11}	cm/s	[31, 32]
	K^+ permeability	$p_{K,rc}$	2.0×10^{-10}	cm/s	[31, 33]
	Cl^- permeability	$p_{Cl,rc}$	2.0×10^{-8}	cm/s	[33, 34]
	HCO_3^- permeability	$p_{HCO3,rc}$	2.1×10^{-4}	cm/s	[35]
	Urea permeability	$p_{U,rc}$	3.0×10^{-4}	cm/s	[36]
	Creatinine permeability	$p_{Cr,rc}$	1.7×10^{-8}	cm/s	[37]
	Na^+ equilibrium ratio	$\beta_{Na,rc}$	0.13	–	Calculated
	K^+ equilibrium ratio	$\beta_{K,rc}$	28.33	–	Calculated
	Cl^- equilibrium ratio	$\beta_{Cl,rc}$	0.70	–	Calculated
	HCO_3^- equilibrium ratio	$\beta_{HCO3,rc}$	0.81	–	Calculated
	Reflection coefficient for ions	$\sigma_{ions,rc}$	1.0	–	Assumed
	Urea reflection coefficient	$\sigma_{U,rc}$	0.7	–	[22]
	Creatinine reflection coefficient	$\sigma_{Cr,rc}$	0.7		(=urea)
	Water permeability	$p_{w,rc}$	7×10^{-3}	cm/s	[38]
	Water transfer coefficient	$K_{w,rc}$	7.56×10^{-5}	L^2/min/mmol/m^2	Calculated

assuming normal concentrations of Na^+, K^+, Cl^- and HCO_3^- in plasma water (149.8, 4.8, 111.4 and 25.7 mmol/L, respectively [146]) and in RBC water (19, 136, 78 and 20.8 mmol/L, respectively [147, 148]).

The parameter Lp (the whole-body capillary hydraulic conductivity) was given the value 4.5 mL/min/mmHg as a compromise between values found by Schneditz et al. [29] and Yashiro et al. [30] in dialysis patients.

The capillary permeability-surface products (PS) for albumin and globulins were calculated from the initial steady-state conditions to obtain the transcapillary escape rate (TER) of 5% of total plasma albumin per hour and 3% of total plasma globulins per hour [149] (TER for IgG was assumed to be representative for all globulins on the basis of a similar average molecular weight of 170 kDa [15, 127, 149]).

2.5 Initial Conditions

The initial conditions of the system were established in two steps. Firstly, the steady-state conditions of the system were defined for a normal, healthy individual with no fistula, thus representing a physiological model of an average healthy subject (potentially useful for other studies). Secondly, a special procedure (described in detail in Sect. 2.5.2) was employed to reach a new steady state of the system representing a typical patient just before dialysis.

2.5.1 Healthy Subject

An active, but untrained 70 kg, 1.75 m tall, mature male individual in the supine position was assumed as the reference healthy subject. The total body water (TBW, in L) was calculated using the anthropometric formula for men by Hume [150]:

$$\mathrm{TBW} = 19.5 \times H + 0.3 \times \mathrm{BW} - 14.0 \tag{2.83}$$

where H is height (in m) and BW is total body weight (in kg).

Based on data presented in [93], 35.7% of TBW was assigned to the extravascular extracellular fluid, i.e. the interstitial fluid, and 53.7% of TBW was assigned to the extravascular intracellular fluid (see Table 2.10). As described in Sect. 2.3.1, the above water volumes correspond to the extracellular and intracellular water content of skeletal muscles, skin, brain, viscera, adipose tissue and bone marrow as well as the water content of bones, connective tissue and transcellular fluids. The remaining 10.6% of TBW was assigned to blood water (i.e. the sum of plasma water and RBC water) [93]. The percentage values given above are slightly different for women (39.5%, 53.7% and 10.8%, respectively [93]). The total volume of the extravascular intracellular fluid (V_{ic}) was calculated assuming the intracellular fluid water fraction $F_{ic,0} = 0.7$ [151]. The initial (normal) water fraction of the interstitial fluid ($F_{is,0}$), used for calculating the total volume of the interstitial fluid (V_{is}), was computed from the steady-state conditions (see below).

Central haematocrit (HCT_C) was assumed at a normal value of 47% [118]. The ratio of whole-body HCT to central HCT (the F-cells ratio [82]) was taken as 0.9, as reported in the literature [82, 83] (assuming no oedema, which decreases this ratio slightly [83]). Assuming plasma water fraction 0.94 and RBC water fraction 0.72 [21], the total blood volume was calculated as 5.15 L (see Table 2.10) or 73.5 mL/kg body weight, which corresponds to the typically reported values of 70–75 mL/kg [137].

The initial (normal) pressures and initial blood distribution across the system (i.e. normal volumes of each blood compartment) were taken from the literature as a compromise between different sources in order to obtain realistic average levels (see Tables 2.6 and 2.7). The initial (normal) vascular resistances (R_i) were calculated from the pressure differences between the compartments (Eq. 2.2), taking the blood flow across the whole system equal to the normal cardiac output (5.25 L/min, assuming a normal stroke volume of 70 mL and a normal heart rate of 75 bpm [152]).

In order to satisfy the assumed ratio of whole-body HCT to central HCT, the tube HCT in the systemic capillaries was given a considerably lower value (21.3%), which correlates with the literature data [6] (even lower HCT values were reported for the capillaries [6, 7]). The initial tube HCT in other compartments was assigned based on the ratio of high-HCT blood volume and low-HCT blood volume

Table 2.6 Normal mean blood pressures across the cardiovascular system [mm Hg] (some values correspond to two or more compartments combined)

<table>
<tr><th>CVS compartment</th><th>Symbol</th><th>[5]</th><th>[14]</th><th>[40]</th><th>[41]</th><th>[14]</th><th>[13]</th><th>This study</th></tr>
<tr><td>Aorta and large arteries</td><td>P_{la}</td><td rowspan="2">100</td><td rowspan="2">100</td><td>95</td><td>100</td><td>100</td><td>95</td><td>95</td></tr>
<tr><td>Small arteries and arterioles</td><td>P_{sa}</td><td>90</td><td>90</td><td>98</td><td>90</td><td>92</td></tr>
<tr><td>Systemic capillaries</td><td>P_{sc}</td><td>–</td><td>–</td><td>–</td><td>35</td><td>30</td><td>30</td><td>32</td></tr>
<tr><td>Venules and small veins</td><td>P_{sv}</td><td rowspan="2">5</td><td>–</td><td>12</td><td>15</td><td>10</td><td>12</td><td>12</td></tr>
<tr><td>Large veins and vena cavae</td><td>P_{lv}</td><td>–</td><td>10</td><td>2</td><td>2</td><td>3</td><td>4</td></tr>
<tr><td>Right heart (right atrium)</td><td>P_{rat}</td><td>4</td><td>0</td><td>–</td><td>0</td><td>1</td><td>2</td><td>1</td></tr>
<tr><td>Pulmonary arteries</td><td>P_{pa}</td><td>17</td><td>16</td><td>–</td><td>–</td><td>15</td><td>–</td><td>16</td></tr>
<tr><td>Pulmonary veins</td><td>P_{pv}</td><td>7</td><td>–</td><td>–</td><td>–</td><td>5</td><td>3–7</td><td>6</td></tr>
<tr><td>Left heart (left atrium)</td><td>P_{lat}</td><td>6.5</td><td>–</td><td>–</td><td>–</td><td>4</td><td>–</td><td>5</td></tr>
</table>

Table 2.7 Blood distribution across the cardiovascular system [%] under steady-state conditions (some values correspond to two or more compartments combined)

<table>
<tr><th>CVS compartment</th><th>[14]</th><th>[42]</th><th>[13]</th><th>[43] (man)</th><th>[43] (dog)</th><th>[44]</th><th>This study</th></tr>
<tr><td>Large systemic arteries</td><td rowspan="2">13</td><td>8.8</td><td rowspan="3">16</td><td rowspan="3">11.4</td><td>8.0</td><td rowspan="3">10–12</td><td>8</td></tr>
<tr><td>Small systemic arteries</td><td>5.9</td><td>3.0</td><td rowspan="2">5</td></tr>
<tr><td>Systemic arterioles</td><td rowspan="2">7</td><td>1.8</td><td>0.4</td></tr>
<tr><td>Systemic capillaries</td><td>5.0</td><td>4</td><td>5.4</td><td>5.2</td><td>4–5</td><td>5</td></tr>
<tr><td>Systemic venules</td><td rowspan="3">64</td><td>9.0</td><td rowspan="3">60</td><td rowspan="3">70</td><td>7.5</td><td rowspan="3">60–68</td><td rowspan="2">26</td></tr>
<tr><td>Small systemic veins</td><td>11.6</td><td>23.9</td></tr>
<tr><td>Large systemic veins</td><td>42.4</td><td>35.2</td><td>39</td></tr>
<tr><td>Pulmonary arteries</td><td rowspan="5">9</td><td>2.8</td><td rowspan="7">20</td><td rowspan="2">2.1</td><td>2.8</td><td rowspan="5">10–12</td><td rowspan="3">6</td></tr>
<tr><td>Pulmonary arterioles</td><td>0.1</td><td>0.5</td></tr>
<tr><td>Pulmonary capillaries</td><td>2.0</td><td>2.6</td><td>4.0</td></tr>
<tr><td>Pulmonary venules</td><td>0.3</td><td rowspan="2">3.5</td><td>0.8</td><td rowspan="2">4</td></tr>
<tr><td>Pulmonary veins</td><td>3.8</td><td>3.7</td></tr>
<tr><td>Right heart</td><td rowspan="2">7</td><td rowspan="2">6.5</td><td rowspan="2">5</td><td rowspan="2">5</td><td rowspan="2">8–11</td><td>3.5</td></tr>
<tr><td>Left heart</td><td>3.5</td></tr>
</table>

approximated from data shown in Table 2.7, as follows: 3:1 for small systemic arteries, 2:1 for small systemic veins, 1:1 for pulmonary arteries and 10:1 for pulmonary veins.

Plasma concentrations of Na^{+}, K^{+}, Cl^{-}, $HCO3^{-}$, Cat^{2+} (ionised fraction of Ca and Mg), urea, creatinine, albumin and total protein were given values from the normal range [118] (see Table 2.8). The initial plasma concentration of globulins was calculated as the difference between the total protein concentration and albumin concentration (assuming that plasma proteins include albumin and globulins only, thus ignoring fibrinogen). The plasma concentration of other anions (An^{2-}) was assigned to obtain plasma electroneutrality (albumin charge was given the value −17 [119], assuming a constant plasma pH equal 7.4 [14], whereas the average charge of

Table 2.8 Initial solute concentrations in the body fluid compartments

Fluid compartment	Solute	Charge	Conc.	Unit	Reference
Plasma	Na^+	+1	140	mmol/L	[118]
	K^+	+1	4.3		
	Cl^-	−1	103		
	HCO_3^-	−1	26		
	Cat^{2+}	+2	2		
	An^{2-}	−2	3.7		Calculated
	Urea	0	6		[118]
	Creatinine	0	0.1		[118]
	Total protein	N/A	7	g/dL	[118]
	Albumin	−17 [119]	4.1		[118]
	Globulins	−11 [120]	2.9		Calculated
Interstitium (extravascular extracellular fluid)	Na^+	+1	145.35	mmol/L	Calculated
	K^+	+1	4.46		
	Cl^-	−1	112.27		
	HCO_3^-	−1	28.34		
	Cat^{2+}	+2	2.03		
	An^{2-}	−2	4.08		
	urea	0	6.38		
	Creatinine	0	0.11		
	Total protein	N/A	2.6	g/dL	Calculated
	Albumin	−17 [119]	1.8		
	Globulins	−11 [120]	0.8		
Tissue cells (extravascular intracellular fluid)	Na^+	+1	8.92	mmol/L	Calculated
	K^+	+1	91.22		
	Cl^-	−1	2.01		
	HCO_3^-	−1	7.34		
	Urea	0	4.47		
	Creatinine	0	0.07		
	Cat^{2+}	+2	33.13		
	An^{2-}	−2	57.20		
	Proteins	−8 [14]	16	g/dL	[14]
Erythrocytes	Na^+	+1	13.60	mmol/L	Calculated
	K^+	+1	93.32		
	Cl^-	−1	55.24		
	HCO_3^-	−1	16.12		
	Urea	0	4.60		
	Creatinine	0	0.08		
	Cat^{2+}	+2	16.76		
	An^{2-}	−2	12.02		
	Proteins (HGB)	−9 [121]	34.04	g/dL	Calculated [122]

globulins was calculated as -11 to obtain the total net charge of all plasma proteins equal to 12 mEq/L [120]).

The initial concentrations of all solutes in the interstitial fluid (see Table 2.8), the initial water fraction of the interstitial fluid ($F_{is,0} \approx 0.98$), the transcapillary Gibbs-Donnan ratio for monovalent cations ($\alpha_{cap} \approx 0.98$) and the normal mean capillary hydraulic pressure ($P_{sc,mean} \approx 17.8$ mm Hg) were calculated (1) to ensure equilibrium between the convective-diffusive mass inflow of solutes from plasma to the interstitium and the convective outflow of solutes from the interstitium with the lymph flow (assumed 8 L/day [93, 115]), (2) to ensure a volumetric balance of transcapillary fluid filtration and the lymph flow and (3) to ensure electroneutrality of the interstitial fluid within the space accessible for proteins (interstitial proteins were given the same charge as plasma proteins). This procedure was realised by minimising the sum of squared deviations from the above steady-state conditions using the Matlab built-in function *fminsearch*, with the starting values of the searched parameters assumed or precalculated close to the final values.

Following the above procedure, in order to obtain electroneutrality of the interstitial space excluded to proteins, a fixed negative charge (5.1 mEq/L) was added to this space representing the charge of the extracellular matrix elements (mainly glycosaminoglycans) [114].

Note that with all solute concentrations calculated as described above, the system is not exactly at steady state due to transcapillary water and solute exchange and the lymph flow, which change slightly the solute concentrations in the capillaries and large veins and hence change the concentrations across the whole system (this effect is, however, negligible and is quickly resolved by the model at the start of simulations).

For both tissue cells and RBCs, the intracellular concentrations of all small solutes (except other cations and anions) were calculated to obtain chemical steady-state equilibrium across the cellular membrane (accounting for the active transport mechanisms). The initial concentration of proteins in the tissue cells was given a normal physiological value of 16 g/dL [14] (the molar concentration was calculated assuming the average molecular weight of intracellular proteins of 30,000 g/mol [153]). The average charge of intracellular proteins was calculated assuming their intracellular concentration of 40 mEq/L [14]. Initial molar concentration of HGB in RBCs was calculated from the assumed normal haemoglobin concentration in the venous blood (16 g/dL) [118], its molecular weight of 68,000 g/mol [15] and the assumed normal value of central HCT. The average haemoglobin charge -9 was taken from [121], assuming a constant pH of RBCs equal to 7.4 [121]. The intracellular concentrations of other cations and anions (both in tissue cells and RBCs) were calculated to obtain cellular electroneutrality and osmotic equilibrium across the cellular membrane (see Table 2.8).

2.5.2 *Patient Before Dialysis*

Once the modelled system represents the steady-state conditions of a reference healthy subject, a special procedure is carried out to simulate a reference patient before dialysis. This procedure involves:

- Addition of a certain amount of water to the system to represent fluid overload
- Adjusting the amounts of selected solutes in the system to represent typical plasma solute concentrations seen in dialysis patients
- Removal of a certain amount of erythrocytes from the cardiovascular system to reduce the haematocrit (low haematocrit is typical for patients with chronic kidney disease [57, 58])
- 'Opening' of the arteriovenous fistula (previously 'closed' for simulating the healthy subject)

In order to simulate the above changes the following steps are performed. Firstly, the plasma concentrations of selected ions and small solutes are changed to represent the typical pre-dialysis levels seen in chronic renal failure patients, i.e. low bicarbonate level, high potassium level and elevated levels of urea and creatinine (see Table 2.9). Along with the above, the concentrations of ions and small solutes in other fluid compartments, i.e. in the interstitial fluid and in the intracellular fluid of tissue cells and erythrocytes, are recalculated (as described for the healthy subject) to keep the whole system at the steady state. The above changes of solute concentrations essentially do not affect water or protein distribution across the system, and hence the assigned volumes of individual compartments and all parameters calculated from the steady-state conditions for the healthy subject are either unchanged or are subject to negligibly small changes related mainly to the accuracy of computations.

Secondly, the excess fluid (water and solutes) is added to the system. It was assumed that the patient has 3 L of fluid overload before dialysis [125, 156]. Given that dialysis patients accumulate the excess fluid almost exclusively in the extracellular space [157], this fluid is added to the interstitial compartment, from which it can distribute to other fluid spaces according to the previously described transport processes. In terms of composition, the added fluid is identical to the interstitial fluid, except for the lack of proteins (it was assumed that the reference patient does not accumulate proteins between dialyses, i.e. the protein catabolism rate is equal to the rate of protein intake) and a slightly different concentration of other anions to

Table 2.9 Assumed pre-dialysis plasma concentrations of ions and small solutes [154, 155]

Sodium	140	mmol/L
Potassium	5	mmol/L
Chloride	103	mmol/L
Bicarbonate	22	mmol/L
Other cations (Ca^{2+}, Mg^{2+})	2	mmol/L
Urea	27	mmol/L
Creatinine	1	mmol/L

keep the added fluid electroneutral. The excess fluid is added to the system gradually, and the model simulation is run for 72 hours to represent a typical 3-day period before the first dialysis in the week (during subsequent 2-day interdialytic periods, patients typically accumulate adequately less fluid). During this simulation the generation of urea and creatinine are temporarily set to zero in order not to add to the system any extra solutes. Given the relatively long simulation time, following the above procedure the modelled system is in an almost steady state, which corresponds to the expectation that patients before dialysis are in a physiological steady state or close to such).

Simultaneously to adding the excess fluid, erythrocytes are removed from the vascular system. To minimise the instantaneous influence on the system, the erythrocytes are removed from the large veins compartment being the largest vascular compartment. In order to maintain the constant blood volume, the volume of removed erythrocytes is replaced by equivalent volume of plasma (with solute concentrations as in plasma). Since the water fractions of RBC fluid is lower than that of plasma, this step results in a slight increase in the total body water (beyond the increase associated with the added water), as can be seen in Table 2.10. The amount of erythrocytes to be removed from the system is determined numerically (within the numerical procedure described below), so that in the new steady state the central haematocrit equals 35%, as seen in dialysis patients [53–55].

The resistance of the arteriovenous fistula (assumed constant) is also found numerically, so that in the new steady state before dialysis, the blood flow through the fistula equals 950 mL/min [158] (‘opening’ of the fistula creates a new path for

Table 2.10 Fluid status assumed for the healthy subject and simulated for the reference patient before dialysis

Parameter	Symbol	Unit	Normal value	References	Pre-dialysis value	References
Total body water	TBW	L	41.13	Eq. 2.83	44.25	Simulated
Extravascular intracellular water	$V_{ic,w}$	L	22.08	Calculated	22.08	
Interstitial water	$V_{is,w}$	L	14.82	Calculated	17.86	
Total blood volume	TBV	L	5.15	Calculated	5.11	
Plasma volume	PV	L	2.97	Calculated	3.50	
RBC volume	RCV	L	2.18	Calculated	1.61	
Plasma water	PV_w	L	2.79	Calculated	3.30	
RBC water	RCV_w	L	1.57	Calculated	1.16	
Central haematocrit	HCT_C	%	47.0	[118]	35.0	[53–55]
Capillary haematocrit	HCT_{sc}	%	21.3	Calculated	15.9	Simulated
Plasma water fraction	F_{pl}	–	0.94	[21]	~0.945	
RBC water fraction	F_{rc}	–	0.72	[21]	~0.720	
Intracellular fluid water fraction	$F_{ic,0}$	–	0.70	[151]	~0.700	
Interstitial fluid water fraction	$F_{is,0}$	–	~0.98	Computed	~0.985	

the blood flow in the system, which reduces the total vascular resistance and increases the cardiac output as indicated by several authors [159, 160]).

With the changes applied to the system as described above, the pre-dialysis steady state of the system depends on the final levels of all solutes, especially proteins (with different protein levels, there will be different oncotic pressure difference acting across the capillary walls and cellular membranes), as well as on the assumed parameters describing the modelled system (e.g. the compliance of the interstitial space or the vascular compartments, sensitivity of stroke volume to changes in preload, etc.). Hence, depending on the amount of water and solutes added to or removed from the system, the new steady state of the system (i.e. the distribution of water and solutes, vascular pressures, etc.) can be different. In particular, the arterial pressure and cardiac output can be markedly different from the expected values. In a real patient, several regulatory mechanisms are activated in response to the fluid overload via signals from different receptors across the whole system (baroreceptors, chemoreceptors and osmoreceptors). These include a range of neurohormonal mechanisms that regulate the arterial pressure and cardiac output in the long term. The proposed model does not include explicitly any of such long-term mechanisms, which are beyond the scope of this study. However, in order to better represent the initial pre-dialysis state of the analysed patient, the modelled system can be relatively easily adjusted by changing the initial pre-dialysis resistance of small arteries compartment and the level of cardiac contractility (both being later regulated in the model by the baroreflex) to reach the desired pre-dialysis arterial blood pressure and cardiac output (two variables that can be easily or relatively easily measured or estimated in real patients). Such adjustments of the vascular resistance and cardiac strength can reflect the long-term adaptation of the cardiovascular system to the presence of arteriovenous fistula, fluid overload and anaemia (reduced haematocrit implies a lower blood viscosity and lower resistance to the blood flow according to Eq. 2.9) and hence can reflect the effects of the long-term mechanisms of arterial blood pressure control or cardiac remodelling, such as left ventricular hypertrophy [161] (for simplicity, the adjustment of heart contractility in the model is applied to both ventricles equally). Such adjustments can also reflect a natural variability of cardiac strength or arterial pressure in different patients, not necessarily related to the presence of fistula or fluid overload. In the present study, these variable parameters of the cardiovascular system were adjusted so that the pre-dialysis transmural arterial BP remains as assumed for the healthy subject, i.e. 96 mmHg (thus assuming that the long-term regulatory mechanisms efficiently maintain the arterial pressure at the normal level, despite fluid overload and despite the fistula-induced and anaemia-induced reduction in the total vascular resistance), and the pre-dialysis cardiac output (q) equals 6.2 L/min as calculated from the assumed access blood flow rate, Q_{ac} (in L/min), using the following formula [158]:

$$q = 0.564 \times Q_{ac}^3 - 2.1964 \times Q_{ac}^2 + 3.8853 \times Q_{ac} + 4.0145 \quad (2.84)$$

The above cardiac output value is almost 20% higher than the normal cardiac output (75 mL/min/kg body weight [137]), which is expected in the system with an arteriovenous fistula, as indicated by several authors [159, 160, 162]). The required adjustments were relatively moderate – an increase of arterioles resistance by 0.12 mmHg∗s/mL (the original resistance was 0.69 mmHg∗s/mL) and an increase of heart contractility by 7% (the parameter E from Eq. (2.11) increased from 1.0 to 1.07). These adjustments may reflect the increased sympathetic activity seen in patients following the creation of arteriovenous fistula [162, 163]. A normal heart rate (75 bpm) was assumed for the pre-dialysis state, but the heart rate can also be directly adjusted in the model, and hence the adjustment of cardiac contractility can be performed with a different heart rate, if required.

Given that the mechanisms regulating arterial pressure in the long term are modelled indirectly, during the above procedure, the baroreflex mechanisms in the model are 'switched off'. It is then assumed that the baroreflex mechanisms adapt to slow, long-term changes in the system and reset around the new pre-dialysis steady-state vascular pressures, which are treated as new normal (reference) pressures for the baroreflex mechanisms, once they are activated during the simulation of dialysis. The above approach is of course a simplification, since most likely the baroreflex mechanisms would not be fully reset to the new operating conditions [164–166]. However, this study is focused on the short-term action of baroreflex during dialysis itself, and the exact mechanisms of chronic baroreflex resetting [167–170] are beyond the scope of this study (see the discussion on model limitations in Sect. 3.3). Further versions of the model could include a detailed description of the long-term pressure regulatory mechanisms to better describe the pre-dialysis steady-state conditions.

To sum up, the whole procedure of reaching the pre-dialysis steady-state conditions includes adding excess fluid (water with ions and small solutes with concentrations as in the initial interstitial fluid), removing erythrocytes, 'opening' the arteriovenous fistula and, finally, adjusting the initial resistance of small arteries and cardiac contractility. Four parameters describing this procedure, i.e. the required amount of erythrocytes to be removed, the required resistance of arteriovenous fistula, the required adjustment of the resistance of small arteries and the required adjustment of cardiac contractility, are found numerically by minimising the sum of squared deviations between the obtained and desired final values of four system variables, i.e. central haematocrit, access blood flow rate, arterial blood pressure and cardiac output. This optimization procedure was performed using the aforementioned Matlab built-in function *fminsearch*. No constraints of the parameters were used.

With the presented approach, starting from the model of a healthy subject, one can describe the pre-dialysis state of the body with any given plasma solute concentrations, haematocrit level, access blood flow rate, cardiac output and arterial blood pressure – the variables that should sufficiently describe a given patient or a group of patients. If needed, during the above procedure, other parameters of the system (e.g. total body water, the transcapillary escape rate of albumin, the lymph flow rate, etc.) can be adjusted, if data on such parameters is available.

Adding to the system, a protein-free fluid results in a decrease of concentration of albumin and globulins in both plasma and interstitial fluid. In the described simulation, the albumin concentration decreased from 4.1 g/dL to 3.7 g/dL, and the total plasma protein concentration decreased from 7 g/dL to 6.4 g/dL. Similar low plasma protein levels are seen in patients before dialysis [123, 171]. The level of blood haemoglobin decreased from 16 g/dL set for the healthy subject to 11.9 g/dL in the pre-dialysis state (similar or lower values are typically seen in patients before dialysis session [54, 122]). Due to the decreased concentration of plasma proteins, which reduces the plasma oncotic pressure, the total blood volume is slightly reduced for the reference patient before dialysis compared to the healthy subject, as can be seen in Table 2.10 (bear in mind that the changes in plasma volume and RBC volume shown in Table 2.10 result mainly from the procedure of replacing the certain amount of erythrocytes by the equivalent plasma volume, as described earlier).

If needed, during the described procedure, proteins can be added or removed to/from the system in order to reach a given protein level seen in the analysed patient or a group of patients (the haemoglobin level can be very easily adjusted by modifying the initial level set for the healthy subject, given that haemoglobin is assumed not to cross the RBC membrane).

Figure 2.11 shows that the percentage blood distribution among all vascular compartments remains very similar in the simulated fluid-overloaded patient compared to the values assumed for the healthy subject. Note a slight increase in relative

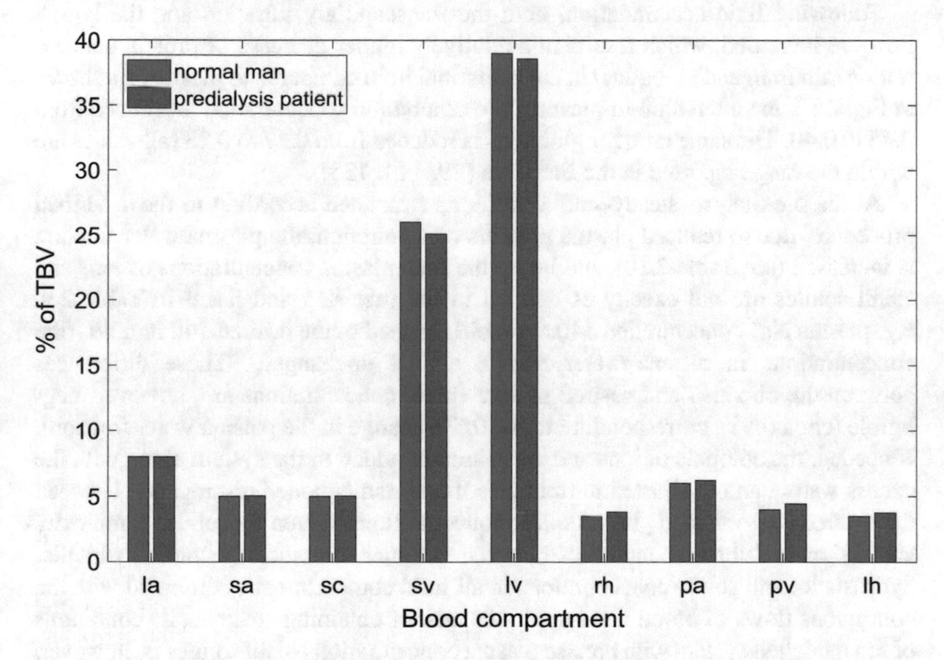

Fig. 2.11 Distribution of blood among the vascular compartments in the healthy subject (assumed) and in the reference patient following 3 L fluid overload (simulated by the model)

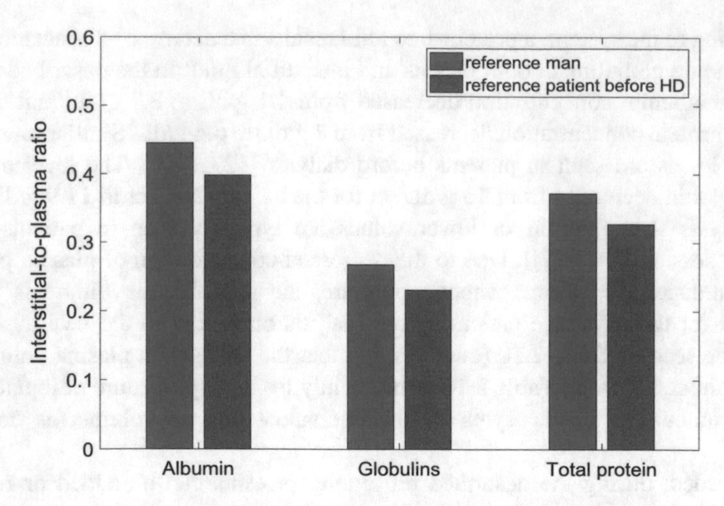

Fig. 2.12 Interstitial-to-plasma ratio of albumin and globulin concentration in the healthy subject and in the reference patient following 3 L fluid overload applied to the model

blood volume in all cardiopulmonary compartments (right and left cardiac compartments as well as pulmonary arterial and venous compartments) reflecting the increased filling pressures of the ventricles and the increased cardiac output (Fig. 2.11).

Following fluid accumulation, both the transcapillary filtration and the lymph flow are increased, which results in a relatively higher decrease of protein concentration (albumin and globulins) in the interstitial fluid compared to plasma. As shown in Fig. 2.12, the interstitial-to-plasma ratio of albumin concentration is reduced from 0.45 to 0.40. The same ratio for globulins is reduced from 0.27 to 0.23 (all values are within the range reported in the literature [29, 114, 125]).

At the pre-dialysis steady-state conditions simulated according to the described procedure, due to reduced plasma proteins concentration, the plasma water fraction is increased (see Table 2.10), and hence the final plasma concentrations of ions and small solutes are not exactly as defined in the first step and listed in Table 2.9, e.g. plasma Na^+ concentration 140.6 mmol/L instead of the desired 140 mmol/L (the concentrations in plasma water remain almost unchanged). These differences between the obtained and desired plasma solute concentrations are, however, negligible (circa 0.5%, corresponding to the 0.5% change in the plasma water fraction). If needed, the amounts of ions and small solutes added to the system along with the excess water can be adjusted to minimise the aforementioned discrepancy between the desired and obtained plasma solute concentrations. Given the model complexity and the mutual interdependencies between volumes, pressures (osmotic/hydraulic/hydrostatic) and solute concentrations in all fluid compartments, combined with the continuous flows of blood and lymph, the task of obtaining steady-state conditions of the modelled system with precise plasma concentrations of all solutes is, however, not easy from the computational point of view. For the description of a typical reference dialysis patient in this study, the used procedure was deemed sufficient.

2.6 Haemodialysis Modelling

A standard 4-hour HD session with a constant ultrafiltration rate and a constant dialysate fluid composition was assumed. The concentrations of Na^+, K^+, Cl^-, $HCO3^-$ and Cat^{2+} in the dialysate fluid were fixed at the typical level with respect to assumed plasma water concentrations (see Table 2.11) [172]. Profiling of the dialysis parameters, i.e. custom intradialytic variations of the ultrafiltration rate or Na^+ in the dialysate fluid, can easily be simulated, if needed.

For urea clearance the expected in vivo clearance was used based on data presented in [172] assuming a high-efficiency dialyzer (K_0A = 1000 mL/min) and the blood flow rate of 300 mL/min. Creatinine clearance was assumed as 80% of urea clearance [172]. The dialysances of all ions (including unspecified cations) were assumed equal to urea clearance, except for $HCO3^-$ dialysance which was assumed as 50% of urea clearance based on data from [173].

Before simulating the HD session, the model accounts for filling of the extracorporeal circuit (i.e. the arterial and venous tubing and the dialyzer) with the patient's blood from the large arterial compartment at the assumed rate of 100 mL/min. With the assumed volume of the extracorporeal circuit (see Table 2.11) filling of the circuit takes in such a case circa 2 minutes.

Table 2.11 Assumed parameters of the haemodialysis treatment

Parameter	Symbol	Value	Unit	Reference
Arterial tubing volume	V_{at}	60	mL	Assumed based on [172]
Venous tubing volume	V_{vt}	60	mL	
Dialyzer volume	V_d	100	mL	
Priming blood flow rate	Q_p	100	mL/min	Assumed
Target blood flow rate	$Q_{b,d}$	300	mL/min	Assumed
Ultrafiltration rate (saline discarded)	Q_{uf}	12.5	mL/min	Calculated
Ultrafiltration rate (saline not discarded)	Q_{uf}	13.4	mL/min	Calculated
Urea clearance	D_U	210	mL/min	[172]
Creatinine clearance	D_{Cr}	168	mL/min	(=0.8D_U) [172]
Na^+ dialysance	D_{Na}	210	mL/min	(=D_U)
K^+ dialysance	D_K	210	mL/min	(=D_U)
Cl^- dialysance	D_{Cl}	210	mL/min	(=D_U)
HCO_3^- dialysance	D_{HCO3}	105	mL/min	(=0.5D_U) [173]
Cat^{2+} dialysance	D_C	210	mL/min	(=D_U)
Na^+ concentration in the dialysate	$c_{Na,d}$	142	mmol/L	Assumed
K^+ concentration in the dialysate	$c_{K,d}$	2	mmol/L	
Cl^- concentration in the dialysate	$c_{Cl,d}$	108	mmol/L	
HCO_3^- concentration in the dialysate	$c_{HCO3,d}$	34	mmol/L	
Cat^{2+} concentration in the dialysate (Mg^{2+}, Ca^{2+})	$c_{C,d}$	2	mmol/L	
Urea concentration in the dialysate	$c_{U,d}$	0	mmol/L	
Creatinine concentration in the dialysate	$c_{Cr,d}$	0	mmol/L	

Two cases are considered in the model: (1) the case when the fluid present in the dialyzer circuit before the session (the priming fluid) is discarded into a drain bag (in such a case, the total volume of blood circulating within the body is reduced) or (2) the case when the priming fluid is infused into the patient (in such a case the total circulating blood volume remains unchanged). The latter case is simulated by infusing the priming solution to the large veins compartment. It is assumed that this solution is the normal saline (154 mmol/L of sodium +154 mmol/L of chloride). When the priming saline is infused into the patient, the dialyzer ultrafiltration has to be appropriately increased in order to remove during dialysis not only the assumed 3 litres of excess fluid but also the volume of the infused saline (equal to the total priming volume of the extracorporeal circuit, see Table 2.11). In both cases, after dialysis the blood from the extracorporeal circuit is infused back into the patient, which obviously increases the vascular volume and blood pressure. This process is not shown in the simulations presented in this study, as it does not have any adverse effects on the cardiovascular system (on the contrary – it increases the volume of blood circulating within the body).

Following the simulation of filling the extracorporeal circuit with the patient's blood, the model simulation is run for 2 minutes during which the dialyzer pump is gradually (assumed linearly) increased up to the target dialyzer blood flow rate (300 mL/minute). During this period the flow of the dialysate fluid in the dialysis machine is assumed to be in the bypass mode with no ultrafiltration and a negligible diffusion of solutes. Following this 'idle' period (during which the medical personnel usually checks the connections of bloodlines and adjusts the dialyzer settings), it is assumed that once the dialyzer blood flow rate reaches the target level, the dialysis machine is switched from the bypass mode into the dialysis mode and the actual haemodialysis with ultrafiltration is started. Note that during this temporary 'idle' circulation of blood through the dialyzer circuit as well as during the preceding filling of the circuit with the blood, the cardiovascular system and the baroreflex mechanisms start to compensate for the changes in blood volume, and hence at the start of the HD simulation, the system is already in a perturbed state.

In the case when the priming saline is discarded into a drain bag, once the patient's blood fills the whole extracorporeal circuit (which is typically detected by a special blood sensor), the dialyzer pump typically stops automatically, so that the venous dialyzer bloodline can be disconnected from the drain bag and connected to the patient. The time needed for this task (most likely below 1 minute) has been neglected in the model but can be added to the simulation, if required. Note also that the exact protocols of dialysis procedures vary from centre to centre.

Given that the dialyzer fibres have usually a diameter of 200–300 μm and are hence much larger than systemic capillaries, the effect of reduced haematocrit in the dialyzer compartment was ignored (the initial haematocrit in the extracorporeal circuit is hence the same as the initial central haematocrit).

2.7 Computational Implementation

The model is based on the set of 330 ordinary differential equations (describing the time changes of all state variables, i.e. the total volumes and water volumes of all 26 compartments, the number of RBCs in each blood compartment, the quantities of the 10 analysed solutes in all 26 compartments and the variables controlled by the baroreflex), circa 800 auxiliary linear or nonlinear algebraic equations and around 150 parameters (the relatively high number of equations results mainly from the repetition of the equations discussed in this chapter for all individual solutes and all compartments analysed in the model). The computer version of the model was implemented in Matlab® (The Mathworks Inc.).

All simulations are performed using the Matlab built-in solver for stiff systems of ordinary differential equations (ode15s). The absolute and relative tolerances of the ODE solver were set to 10^{-3}, except for the sensitivity analysis, for which both the absolute and relative tolerances were set to 10^{-6}.

The auxiliary numerical procedures for establishing the initial steady-state conditions for the healthy subject and the initial pre-dialysis steady-state conditions for the reference patient are based on minimising the sum of squared deviations from the steady-state conditions or deviations from the desired levels of certain variables in the patient before dialysis, as described in Sect. 2.5. As already mentioned, these tasks were realised using the unconstrained Matlab built-in function *fminsearch*, which uses the Nelder-Mead simplex direct search algorithm [174].

Table 2.12 summarises all steps involved in the proposed modelling approach divided into three stages – the definition of the model for the healthy subject, the simulation of fluid overload in the dialysis patient and the simulation of the whole HD procedure.

Table 2.12 Simulation steps and timing used in the proposed modelling approach

No	Stage	Step	Duration of the simulated process
1	Healthy subject (see Sect. 2.5.1)	Defining steady-state conditions for the healthy subject	–
2	Pre-dialysis patient (see Sect. 2.5.2)	Adding excess water and solutes to the body; removing certain amount of erythrocytes from the system	24 hours
3		Reaching steady-state conditions of the fluid-overloaded patient	48 hours
4		Resetting baroreflex mechanisms at the pre-dialysis steady-state conditions	–
5	Dialysis session (see Sect. 2.6)	Filling the extracorporeal circuit with the patient's blood	~2 minutes
6		'Idle' extracorporeal circulation of blood	2 minutes
7		Haemodialysis	4 hours

References

1. Pstras, L., Thomaseth, K., Waniewski, J., Balzani, I., Bellavere, F.: The Valsalva manoeuvre: physiology and clinical examples. Acta Physiol (Oxf.). **217**(2), 103–119 (2016)
2. Pstras, L., Thomaseth, K., Waniewski, J., Balzani, I., Bellavere, F.: Mathematical modelling of cardiovascular response to the Valsalva manoeuvre. Math Med Biol. **34**(2), 261–292 (2017)
3. Pstras, L., Thomaseth, K., Waniewski, J.: Personalised simulation of haemodynamic response to the Valsalva manoeuvre. In: Computational methods in data analysis, ITRIA 2015, pp. 119–134, Warsaw (2015)
4. Ursino, M., Innocenti, M.: Modeling arterial hypotension during hemodialysis. Art Org. **21**(8), 873–890 (1997)
5. Ursino, M., Antonucci, M., Belardinelli, E.: Role of active changes in venous capacity by the carotid baroreflex: analysis with a mathematical model. Am J Physiol Heart Circ Physiol. **267** (6), H2531–H2546 (1994)
6. Schmid-Schoenbein, G., Zweifach, B.: RBC velocity profiles in arterioles and venules of the rabbit omentum. Microvasc Res. **10**(2), 153–164 (1975)
7. Lipowsky, H., Usami, S., Chien, S.: In vivo measurements of "apparent viscosity" and microvessel hematocrit in the mesentery of the cat. Microvasc. Res. **19**(3), 297–319 (1980)
8. Sugihara-Seki, M., Fu, B.: Blood flow and permeability in microvessels. Fluid Dyn Res. **37**, 82–132 (2005)
9. Clark, K., Hippel, T.: Routine and point-of-care testing in hematology: manual and semiautomated methods. In: Hematology: Clinical Principles and Applications, 4th edn, p. 181. Elsevier Saunders, Philadelphia (2011)
10. Schalk, E., Heim, M., Koenigsmann, M., Jentsch-Ullrich, K.: Use of capillary blood count parameters in adults. Vox Sang. **93**(4), 348–353 (2007)
11. Polaschegg, H.: Red blood cell damage from extracorporeal circulation in hemodialysis. Semin Dial. **22**(5), 524–531 (2009)
12. Landis, E., Pappenheimer, J.: Exchange of substances through the capillary walls. In: Hamilton, W., Dow, P. (eds.) Handbook of physiology. Section 2: Circulation 11. American Pysiological Society, Washington, D.C. (1963)
13. Downey, J.: Hemodynamics. In: Essential Medical Phsyiology, 3rd edn. Elsevier Academic Press, Amsterdam (2003)
14. Guyton, A., Hall, J.: Textbook of Medical Physiology, 11th edn. Elsevier Saunders, Philadelphia (2006)
15. Roselli, R., Diller, K.: Biotransport: Principles and Applications. Springer, New York (2011)
16. Pstras, L., Thomaseth, K., Waniewski, J., Balzani, I., Bellavere, F.: Modeling pathological hemodynamic responses to the Valsalva maneuver. J Biomech Eng. **139**(6), 061001-1-9 (2017)
17. Schneditz, D., Platzer, D., Daugirdas, J.: A diffusion-adjusted regional blood flow model to predict solute kinetics during haemodialysis. Nephrol Dial Transplant. **24**, 2218–2224 (2009)
18. Ursino, M., Coli, L., Brighenti, C., Chiari, L., de Pascalis, A., Avanzolini, G.: Prediction of solute kinetics, acid-base status, and blood volume changes during profiled hemodialysis. Ann Biomed Eng. **28**(2), 204–216 (2000)
19. Coli, L., Ursino, M., De Pascalis, A., Brighenti, C., Dalmastri, V., La Manna, G., Isola, E., Cianciolo, G., Patrono, D., Boni, P., Stefoni, S.: Evaluation of intradialytic solute and fluid kinetics. Setting up a predictive mathematical model. Blood Purif. **18**(1), 37–49 (2000)
20. Eloot, S., Torremans, A., De Smet, R., Marescau, B., De Wachter, D., De Peyn, P., Lameire, N., Verdonck, P., Vanholder, R.: Kinetic behavior of urea is different from that of other water-soluble compounds: the case of the guanidino compounds. Kidney Int. **67**(4), 1566–1575 (2005)
21. Sargent, J., Gotch, F.: Principles and biophysics of dialysis. In: Jacobs, C., Kjellstrand, C., Koch, K., Winchester, J. (eds.) Replacement of Renal Function by Dialysis, 4th edn, pp. 34–102. Springer, Dordrecht (1996)

22. Chasan, B., Solomon, A.: Urea reflection coefficient for the human red cell membrane. Biochim Biophys Acta. **821**(1), 56–62 (1985)
23. Mann, H., Stiller, S.: Sodium modeling. Kidney Int Suppl. **76**, S79–S88 (2000)
24. Waniewski, J.: Distributed modeling of diffusive solute transport in peritoneal dialysis. Ann Biomed Eng. **30**, 1181–1195 (2002)
25. Dedrick, R., Flessner, M., Collins, J., Schultz, J.: Is the peritoneum a membrane? ASAIO J. **5**, 1–8 (1982)
26. Chapple, C., Bowen, B., Reed, R., Xie, S., Bert, J.: A model of human microvascular exchange: parameter estimation based on normals and nephrotics. Comput Methods Prog Biomed. **41**(1), 33–54 (1993)
27. Gyenge, C., Bowen, B., Reed, R., Bert, J.: Transport of fluid and solutes in the body I. Formulation of a mathematical model. Am J Physiol Heart Circ Physiol. **277**(3 Pt 2), H1215–H1227 (1999)
28. Wolf, M.: Estimation of whole-body capillary transport parameters from osmotic transient data. Am J Phys. **242**(3), R227–R236 (1982)
29. Schneditz, D., Roob, J., Oswald, M., Pogglitsch, H., Moser, M., Kenner, T., Binswanger, U.: Nature and rate of vascular refilling during hemodialysis and ultrafiltration. Kidney Int. **42**(6), 1425–1433 (1992)
30. Yashiro, M., Hamada, Y., Matsushima, H., Muso, E.: Estimation of filtration coefficients and circulating plasma volume by continuously monitoring hematocrit during hemodialysis. Blood Purif. **20**(6), 569–576 (2002)
31. Hoffman, J.: Active transport of Na+ and K+ by red blood cells. In: Andreoli, T., Fanestil, D., Hoffman, J., Schultz, S. (eds.) Membrane Physiology Part III, pp. 221–234. Springer, New York (1987)
32. LaCelle, P., Rothstein, A.: The passive permeability of the red blood cell to cations. J Gen Physiol. **50**, 171–188 (1966)
33. Parker, J., Dunham, P.: Passive cation transport. In: Agre, P., Parker, J.C. (eds.) Red Blood Cell Membranes: Structure. Function. Clinical Implications, p. 526. Marcel Dekker, Inc., New York (1989)
34. Sachs, J., Knauf, P., Dunham, P.: Transport through red cell membranes. In: Surgenor, D. (ed.) The Red Blood Cell, vol. II, 2nd edn. Academic Press, Inc., London (1975)
35. Chow, E., Crandall, E., Forster, R.: Kinetics of bicarbonate-chloride exchange across the human red blood cell membrane. J Gen Physiol. **68**(6), 633–652 (1976)
36. Brahm, J.: The permeability of red blood cells to chloride, urea and water. J Exp Biol. **216** (Pt 12), 2238–2246 (2013)
37. Langsdorf, L., Zydney, A.: Effect of solution environment on the permeability of red blood cells. Biotechnol Bioeng. **43**(2), 115–121 (1994)
38. Benga, G., Borza, T.: Diffusional water permeability of mammalian red blood cells. Comp Biochem Physiol B Biochem Mol Biol. **112**(4), 653–659 (1995)
39. Snyder, M., Rideout, V.: Computer simulation studies of the venous circulation. IEEE Trans Biomed Eng. **BME-16**(4), 325–334 (1969)
40. Boron, W., Boulpaep, E.: Medical Physiology. Elsevier Saunders, Philadelphia (2012)
41. Patton, K., Thibodeau, G.: Circulation of blood. In: Anatomy and Physiology, 9th edn. Elsevier, St. Louis (2016)
42. Schneck, D.: An outline of cardiovascular structure and function. In: The Biomedical Engineering Handbook, vol. 1, 2nd edn. CRC Press LLC, Boca Raton (2000)
43. Milnor, W.: Hemodynamics. Lippincott William & Wikins, Baltimore (1982)
44. Bell, D.: Overview of the cardiovascular system and hemodynamics. In: Medical Physiology: Principles for Clinical Medicine, 4th edn. Lippincott Williams & Wilkins, Philadelphia (2013)
45. Safar, M., London, G.: Venous system in essential hypertension. In: Arterial and Venous Systems in Essential Hypertension 63, pp. 53–66. Springer, Dordrecht (1987)
46. Heldt, T.: Computational models of cardiovascular response to orthostatic stress. PhD thesis, Massachusetts Institute of Technology (2004)

47. Wong, S., Bredin, S., Krassioukov, A., Taylor, A., Warburton, D.: Effects of training status on arterial compliance in able-bodied persons and persons with spinal cord injury. Spinal Cord. **51**, 278–281 (2013)
48. Tanaka, T., Arakawa, M., Suzuki, T., Gotoh, M., Miyamoto, H., Hirakawa, S.: Compliance of human pulmonary venous system estimated from pulmonary artery wedge pressure tracings--comparison with pulmonary arterial compliance. Jpn Circ J. **50**(2), 127–139 (1986)
49. Hardy, H., Collins, R.: On the pressure-volume relationship in circulatory elements. Med Biol Eng Comput. **20**, 565–570 (1982)
50. Liang, F., Liu, H.: Simulation of hemodynamic responses to the Valsalva maneuver: an integrative computational model of the cardiovascular system and the autonomic nervous system. J Physiol Sci. **56**(1), 45–65 (2006)
51. Sutera, S., Richard, S.: The history of Poiseuille's law. Annu Rev Fluid Mech. **25**, 1–19 (1993)
52. Olufsen, M., Ottesen, J., Tran, H., Ellwein, L., Lipsitz, L., Novak, V.: Blood pressure and blood flow variation during postural change from sitting to standing: model development and validation. J Appl Physiol. **99**(4), 1523–1537 (2005)
53. Metry, G., Adhikarla, R., Schneditz, D., Ronco, C., Levin, N.: Effect of changes in the intravascular volume during hemodialysis on blood viscoelasticity. Indian J Nephrol. **21**(2), 95–100 (2011)
54. Shirazian, S., Rios-Rojas, L., Drakakis, J., Dikkala, S., Dutka, P., Duey, M., Cho, D., Fishbane, S.: The effect of hemodialysis ultrafiltration on changes in whole blood viscosity. Hemodial Int. **16**(3), 342–350 (2012)
55. Dhar, P., Eadon, M., Hallak, P., Munoz, R., Hammes, M.: Whole blood viscosity: effect of hemodialysis treatment and implications for access patency and vascular disease. Clin Hemorheol Microcirc. **51**(4), 265–275 (2012)
56. Stark, H., Schuster, S.: Comparison of various approaches to calculating the optimal hematocrit in vertebrates. J Appl Physiol. **113**(3), 355–367 (1985). (2012)
57. Hsu, C., Bates, D., Kuperman, G., Curhan, G.: Relationship between hematocrit and renal function in men and women. Kidney Int. **59**(2), 725–731 (2001)
58. Fishbane, S., Spinowitz, B.: Update on anemia in ESRD and earlier stages of CKD: core curriculum 2018. Am J Kidney Dis. **71**(3), 423–435 (2018)
59. Lu, K., Clark, J., Ghorbel, F., Ware, D., Bidani, A.: A human cardiopulmonary system model applied to the analysis of the Valsalva maneuver. Am J Physiol Heart Circ Physiol. **281**, H2661–H2679 (2001)
60. Damasiewicz, M., Polkinghorne, K.: Intra-dialytic hypotension and blood volume and blood temperature monitoring. Nephrology. **16**, 13–18 (2011)
61. Fåhraeus, R., Lindqvist, T.: The viscosity of the blood in narrow capillary tubes. Am J Phys. **96**, 562–568 (1931)
62. Secomb, T., Pries, A.: Blood viscosity in microvessels: experiment and theory. C R Phys. **14** (6), 470–478 (2013)
63. Pries, A., Secomb, T.: Microvascular blood viscosity in vivo and the endothelial surface layer. Am J Physiol Heart Circ Physiol. **289**, H2657–H2664 (2005)
64. Ciandrini, A., Cavalcanti, S., Severi, S., Garred, L., Avanzolini, G.: Effects of dialysis technique on acute hypotension: a model-based study. Cardiovasc Eng. **4**(2), 163–171 (2004)
65. Cavalcanti, S., Cavani, S., Ciandrini, A., Avanzolini, G.: Mathematical modeling of arterial pressure response to hemodialysis-induced hypovolemia. Comput Biol Med. **36**(2), 128–144 (2006)
66. Sonnenblick, E., Downing, S.: Afterload as a primary determinant of ventricular performance. Am J Phys. **204**, 604–610 (1963)
67. Rosenblueth, A., Alanis, J., Rubio, R.: Some properties of the mammalian ventricular muscle. Arch Int Physiol Biochim. **67**, 276–293 (1959)
68. Imperial, E., Levy, M.: Outflow resistance as an independent determinant of cardiac performance. Circ Res. **9**, 1148–1155 (1961)

69. Wilcken, D., Charlier, A., Hoffman, J., Gux, A.: Effects of alterations in aortic impedance on the performance of the ventricles. Circ Res. **14**(4), 283–293 (1964)
70. MacGregor, D., Covell, J., Mahler, F., Dilley, R., Ross Jr., J.: Relations between afterload, stroke volume and descending limb of Starling's curve. Am J Phys. **227**, 884–890 (1974)
71. Van Hare, G., Hawkins, J., Schmidt, K., Rudolph, A.: The effects of increasing mean arterial pressure on left ventricular output in newborn lambs. Circ Res. **67**, 78–83 (1990)
72. Cohn, J., Franciosa, J.: Vasodilator therapy of cardiac failure. N Engl J Med. **297**(1), 27–31 (1977)
73. Sunagawa, K., Sagawa, K., Maughan, W.: Ventricular interaction with the loading system. Ann Biomed Eng. **12**, 163–189 (1984)
74. Sarnoff, S., Berglund, E.: Ventricular function. I. Starling's law of the heart studied by means of simultaneous right and left ventricular function curves in the dog. Circulation. **9**(5), 706–718 (1954)
75. Vest, A., Heupler Jr., F.: Afterload. In: Anwaruddin, S., Martin, J., Stephens, J., Askari, A. (eds.) Cardiovascular Hemodynamics: An Introductory Guide, pp. 29–51. Springer Science +Business Media, New York (2013)
76. Ursino, M.: Modelling the interaction among several mechanisms in the short-term arterial pressure control. In: Mathematical Modelling in Medicine, pp. 139–161. IOS Press, Amsterdam (2000)
77. Magosso, E., Biavati, V., Ursino, M.: Role of the baroreflex in cardiovascular instability: a modeling study. Cardiovasc Eng. **1**(2), 101–115 (2001)
78. Ursino, M.: Interaction between carotid barregulation and the pulsating heart: a mathematical model. Am J Phys. **275**(5), H1733–H1747 (1998)
79. Daugirdas, J.: Dialysis hypotension: a hemodynamic analysis. Kidney Int. **39**(2), 233–246 (1991)
80. Fåhraeus, R.: The suspension stability of blood. Physiol Rev. **9**, 241–274 (1929)
81. Fournier, R.: The physical and flow properties of blood. In: Basic Transport Phenomena in Biomedical Engineering, 3rd edn. CRC Press, Boca Raton (2011)
82. Schneditz, D., Ribitsch, W., Schilcher, G., Uhlmann, M., Chait, Y., Stadlbauer, V.: Concordance of absolute and relative plasma volume changes and stability of Fcells in routine hemodialysis. Hemodial Int. **20**, 120–128 (2016)
83. Fudenberg, H., Baldini, M., Mahoney, J., Dameshek, W.: The body hematocrit/venous ratio and the "splenic reservoir". Blood. **17**, 71–82 (1961)
84. Stead Jr., E., Ebert, R.: Relationship of the plasma volume and the cell plasma ratio to the total red cell volume. Am J Phys. **132**, 411–417 (1941)
85. Haynes, R.: Physical basis of the dependence of blood viscosity on tube radius. Am J Phys. **198**, 1193–1200 (1960)
86. Popel, A., Pittman, R.: Mechanics and transport in the microcirculation. In: Schneck, D., Bronzino, J. (eds.) Biomechanics. Principles and Applications. CRC Press LLC, Boca Raton (2003)
87. Mitra, S., Chamney, P., Greenwood, R., Farrington, K.: The relationship between systemic and whole-body hematocrit is not constant during ultrafiltration on hemodialysis. J Am Soc Nephrol. **15**(2), 463–469 (2004)
88. Schneditz, D., Kaufman, A., Polaschegg, H., Levin, N., Daugirdas, J.: Cardiopulmonary recirculation during hemodialysis. Kidney Int. **42**(6), 1450–1456 (1992)
89. Lindsay, R., Blake, P., Malek, P., Posen, G., Martin, B., Bradfield, E.: Hemodialysis access blood flow rates can be measured by a differential conductivity technique and are predictive of access clotting. Am J Kidney Dis. **30**(4), 475–482 (1997)
90. Cingolani, H., Pérez, N., Cingolani, O., Ennis, I.: The Anrep effect: 100 years later. Am J Physiol Heart Circ Physiol. **304**(2), H175–H182 (2013)
91. Sun, Y., Beshara, M., Lucariello, R., Chiaramida, S.: A comprehensive model for right-left heart interaction under the influence of pericardium and baroreflex. Am J Phys. **272**(3 Pt 2), H1499–H1515 (1997)
92. Klahr, S. (ed.): The Kidney and Body Fluids in Health and Disease, 1st edn. Springer Science +Business Media LLC, New York (1983)

93. Bhave, G., Neilson, E.: Body fluid dynamics: back to the future. J Am Soc Nephrol. **22**, 2166–2181 (2011)
94. Wolf, M.: Whole body acid-base and fluid-electrolyte balance: a mathematical model. Am J Physiol Renal Physiol. **305**(8), F1118–F1131 (2013)
95. Debowska, M., Waniewski, J., Lindholm, B.: An integrative description of dialysis adequacy indices for different treatment modalities and schedules of dialysis. Artif Organs. **31**(1), 61–69 (2007)
96. Waniewski, J., Debowska, M., Lindholm, B.: Can the diverse family of dialysis adequacy indices be understood as one integrated system? Blood Purif. **30**(4), 257–265 (2010)
97. Shulman, T., Heidenheim, A., Kianfar, C., Shulman, S., Lindsay, R.: Preserving central blood volume: changes in body fluid compartments during hemodialysis. ASAIO J. **47**(6), 615–618 (2001)
98. Depner, T.: Multicompartment models. In: Prescribing Hemodialysis. A Guide to Urea Modeling. Kluwer Academic Publishers, Boston (2002)
99. Waniewski, J.: Mathematical modeling of fluid and solute transport in hemodialysis and peritoneal dialysis. J Membr Sci. **274**, 24–37 (2006)
100. Waniewski, J.: Transport across permselective membrane. In: Theoretical Foundations for Modelling of Membrane Transport in Medicine and Biomedical Engineering, pp. 24–27. Institute of Computer Science, Polish Academy of Sciences, Warsaw (2015)
101. Manery, J.: Water and electrolyte metabolism. Physiol Rev. **34**(2), 334–417 (1954)
102. Swan, R., Feinstein, H., Madisso, H., Plinke, A., Sharma, H.: Distribution of sulfate ion across semi-permeable membranes. J Clin Invest. **35**, 607–610 (1956)
103. Rippe, B., Venturoli, D., Simonsen, O., de Arteaga, J.: Fluid and electrolyte transport across the peritoneal membrane during CAPD according to the three-pore model. Perit Dial Int. **24** (1), 10–27 (2004)
104. Michel, C., Curry, F.: Microvascular permeability. Physiol Rev. **79**(3), 703–761 (1999)
105. Bert, J., Gyenge, C., Bowen, B., Reed, R., Lund, T.: A model of fluid and solute exchange in the human: validation and implications. Acta Physiol Scand. **170**(3), 201–209 (2000)
106. Starling, E.: On the absorption of fluids from the connective tissue spaces. J Physiol. **19**, 312–326 (1896)
107. Landis, E.: Micro-injection studies of capillary blood pressure in human skin. Heart. **30**, 209–228 (1930)
108. Levick, J.: Capillary filtration-absorption balance reconsidered in light of dynamic extravascular factors. Exp Physiol. **76**(6), 825–857 (1991)
109. Levick, J., Michel, C.: Microvascular fluid exchange and the revised Starling principle. Cardiovasc Res. **87**, 198–210 (2010)
110. Johnson, P.: Effect of venous pressure on mean capillary pressure and vascular resistance in the intestine. Circ Res. **16**, 294–300 (1965)
111. Järhult, J., Mellander, S.: Autoregulation of capillary hydrostatic pressure in skeletal muscle during regional arterial hypo- and hypertension. Acta Physiol Scand. **91**(1), 32–41 (1974)
112. Björnberg, J.: Forces involved in transcapillary fluid movement in exercising cat skeletal muscle. Acta Physiol Scand. **140**(2), 221–236 (1990)
113. Mahy, I., Tooke, J., Shore, A.: Capillary pressure during and after incremental venous pressure elevation in man. J Physiol. **485**(Pt 1), 213–219 (1995)
114. Aukland, K., Reed, R.: Interstitial-lymphatic mechanisms in the control of extracellular fluid volume. Physiol Rev. **73**(1), 1–78 (1993)
115. Renkin, E.: Some consequences of capillary permeability to macromolecules: Starling's hypothesis reconsidered. Am J Phys. **250**(5 Pt 2), H706–H710 (1986)
116. Xie, S., Reed, R., Bowen, B., Bert, J.: A model of human microvascular exchange. Microvasc Res. **49**(2), 141–162 (1995)
117. Nitta, S., Ohnuki, T., Ohkuda, K., Nakada, T., Staub, N.: The corrected protein equation to estimate plasma colloid osmotic pressure and its development on a nomogram. Tohoku J Exp Med. **135**(1), 43–49 (1981)

118. Lagua, R., Claudio, V.: Appendix 30. Reference values for normal blood constituents. In: Nutrition and Diet Therapy Reference Dictionary, 4th edn, pp. 430–436. Chapman & Hall, New York (1996)
119. Fogh-Andersen, N., Bjerrum, P., Siggaard-Andersen, O.: Ionic binding, net charge, and Donnan effect of human serum albumin as a function of pH. Clin Chem. **39**(1), 48–52 (1993)
120. Figge, J., Rossing, T., Fencl, V.: The role of serum proteins in acid-base equilibria. J Lab Clin Med. **117**(6), 453–467 (1991)
121. Freedman, J., Hoffman, J.: Ionic and osmotic equilibria of human red blood cells treated with nystatin. J Gen Physiol. **74**(2), 157–185 (1970)
122. Burton, J., Jefferies, H., Selby, N., McIntyre, C.: Hemodialysis-induced cardiac injury: determinants and associated outcomes. Clin J Am Soc Nephrol. **4**(5), 914–920 (2009)
123. Ookawara, S., Sato, H., Takeda, H., Tabei, K.: Method for approximating colloid osmotic pressure in long-term hemodialysis patients. Ther Apher Dial. **18**(2), 202–207 (2014)
124. Adamson, R., Lenz, J., Zhang, X., Adamson, G., Weinbaum, S., Curry, F.: Oncotic pressures opposing filtration across non-fenestrated rat microvessels. J Physiol. **557**, 889–908 (2004)
125. Pietribiasi, M., Waniewski, J., Załuska, A., Załuska, W., Lindholm, B.: Modelling transcapillary transport of fluid and proteins in hemodialysis patients. PLoS One. **11**(8), e0159748 (2016)
126. Stachowska-Pietka, J., Waniewski, J., Flessner, M., Lindholm, B.: Concomitant bidirectional transport during peritoneal dialysis can be explained by a structured interstitium. Am J Physiol Heart Circ Physiol. **310**(11), H1501–H1511 (2016)
127. Feher, J.: Quantitative Human Physiology. An Introduction. Academic Press/Elsevier, Amsterdam (2012)
128. Parker, J., Falgout, H., Parker, R., Granger, D., Taylor, A.: The effect of fluid volume loading on exclusion of interstitial albumin and lymph flow in the dog lung. Circ Res. **45**(4), 440–450 (1979)
129. Rippe, B., Haraldsson, B.: Transport of macromolecules across microvascular walls: the two-pore theory. Physiol Rev. **74**(1), 163–219 (1994)
130. Maughan, D., Godt, R.: Protein osmotic pressure and the state of water in frog myoplasm. Biophys J. **80**(1), 435–442 (2001)
131. Van der Goot, F., Podevin, R., Corman, B.: Water permeabilities and salt reflection coefficients of luminal, basolateral and intracellular membrane vesicles isolated from rabbit kidney proximal tubule. Biochim Biophys Acta. **986**(2), 332–340 (1989)
132. Fischer, H., Polikarpov, I., Craievich, A.: Average protein density is a molecular-weight-dependent function. Protein Sci. **13**(10), 2825–2828 (2004)
133. London, G., Safar, M.: Venous compliance in essential hypertension. In: Safar, M., London, G., Simon, A., Weiss, Y. (eds.) Arterial and Venous Systems in Essential Hypertension 63, pp. 53–66. Martinus Nijhoff Publishers, Dordrecht (1987)
134. Spaan, J.: Coronary diastolic pressure-flow relation and zero flow pressure explained on the basis of intramyocardial compliance. Circ Res. **56**(3), 293–309 (1985)
135. Mangano, D., Van Dyke, D., Ellis, R.: The effect of increasing preload on ventricular output and ejection in man. Limitations of the Frank-Starling mechanism. Circulation. **62**(3), 535–541 (1980)
136. Vatner, S., Boettcher, D.: Regulation of cardiac output by stroke volume and heart rate in conscious dogs. Circ Res. **42**(4), 557–561 (1978)
137. Rothe, C.: Reflex control of veins and vascular capacitance. Physiol Rev. **63**(4), 1281–1342 (1983)
138. Herman, I.: Physics of the Human Body. Springer, Berlin (2007)
139. Angell James, J., de Burgh Daly, M.: Effects of graded pulsatile pressure on the reflex vasomotor responses elicited by changes of mean pressure in the perfused carotid sinus-aortic arch regions of the dog. J Physiol. **214**(1), 51–64 (1971)
140. Chen, H., Chai, C., Tung, C., Chen, H.: Modulation of the carotid baroreflex function during volume expansion. Am J Physiol Heart Circ Physiol. **237**(2), H153–H158 (1979)

141. Mancia, G., Mark, A.: Arterial baroreflexes in humans. In: Shepherd, J., Abboud, F., Geiger, S. (eds.) Handbook of Physiology. Sect. 2 The Cardiovascular System. III, pp. 755–793. The American Physiological Society, Bethesda (1983)
142. Potts, J., Hatanaka, T., Shoukas, A.: Effect of arterial compliance on carotid sinus baroreceptor reflex control of the circulation. Am J Physiol Heart Circ Physiol. **270**(3 Pt 2), H988–H1000 (1996)
143. Schmidt, R., Kumada, M., Sagawa, K.: Cardiovascular responses to various pulsatile pressures in the carotid sinus. Am J Phys. **223**(1), 1–7 (1972)
144. Shoukas, A., Sagawa, K.: Control of total systemic vascular capacity by the carotid sinus baroreceptor reflex. Circ Res. **33**(1), 22–33 (1973)
145. TenVoorde, B., Kingma, R.: A baroreflex model of short term blood pressure and heart rate variability. In: Ottesen, J., Danielsen, M. (eds.) Mathematical Modelling in Medicine, p. 183. IOS Press, Amsterdam (2000)
146. Oh, M., Briefel, G.: Evaluation of renal function, water, electrolytes, and acid-base balance. In: McPherson, R., Pincus, M. (eds.) Henry's Clinical Diagnosis and Management by Laboratory Methods, 23rd edn, p. 164. Elsevier, St. Louis (2017)
147. Bernstein, R.: Potassium and sodium balance in mammalian red cells. Science. **120**(3116), 459–460 (1954)
148. Kam, P., Power, I.: Acid-base physiology. In: Principles of Physiology for the Anaesthetist, 3rd edn. CRC Press, Boca Raton (2015)
149. Rossing, N.: Intra- and extravascular distribution of albumin and immunoglobulin in man. Lymphology. **11**, 138–142 (1978)
150. Hume, R., Weyers, E.: Relationship between total body water and surface area in normal and obese subjects. J Clin Pathol. **24**(3), 234–236 (1971)
151. Luby-Phelbs, K.: Cytoarchitecture and physical properties of cytoplasm: volume, viscosity, diffusion, intracellular surface area. Int Rev Cytol. **192**, 189–221 (1999)
152. Elad, D., Einav, S.: Physical and flow properties of blood. In: Biomedical engineering and design handbook, vol. 1, 2nd edn. New York City, NY. McGraw-Hill (2009)
153. Philips, R., Kondev, J., Theriot, J., Garcia, H.: Why: biology by the numbers. In: Physical Biology of the Cell, 2nd edn. Garland Science, London (2013)
154. Kyriazis, J., Kalogeropoulou, K., Bilirakis, L., Smirnioudis, N., Pikounis, V., Stamatiadis, D., Liolia, E.: Dialysate magnesium level and blood pressure. Kidney Int. **66**(3), 1221–1231 (2004)
155. Berta, E., Erdei, A., Cseke, B., Gazdag, A., Paragh, G., Balla, J., Polgar, P., Nagy, E., Bodor, M.: Evaluation of the metabolic changes during hemodialysis by signal averaged ECG. Pharmazie. **67**(5), 380–383 (2012)
156. Pietribiasi, M., Waniewski, J., Wojcik-Zaluska, A., Zaluska, W., Lindholm, B.: Model of fluid and solute shifts during hemodialysis with active transport of sodium and potassium. PLoS One. **13**(12), e0209553 (2018)
157. Fisch, B., Spiegel, D.: Assessment of excess fluid distribution in chronic hemodialysis patients using bioimpedance spectroscopy. Kidney Int. **49**(4), 1105–1109 (1996)
158. Basile, C., Lomonte, C., Vernaglione, L., Casucci, F., Antonelli, M., Losurdo, N.: The relationship between the flow of arteriovenous fistula and cardiac output in haemodialysis patients. Nephrol Dial Transplant. **23**(1), 282–287 (2008)
159. Dundon, B., Torpey, K., Nelson, A., Wong, D., Duncan, R., Meredith, I., Faull, R., Worthley, S., Worthley, M.: The deleterious effects of arteriovenous fistula-creation on the cardiovascular system: a longitudinal magnetic resonance imaging study. Int J Nephrol Renov Dis. **16**(7), 337–345 (2014)
160. Korsheed, S., Eldehni, M., John, S., Fluck, R., McIntyre, C.: Effects of arteriovenous fistula formation on arterial stiffness and cardiovascular performance and function. Nephrol Dial Transplant. **26**(10), 3296–3302 (2011)
161. London, G.: Left ventricular alterations and end-stage renal disease. Nephrol Dial Transplant. **17**(Suppl 1), 29–36 (2002)

162. Alkhouli, M., Sandhu, P., Boobes, K., Hatahet, K., Raza, F., Boobes, Y.: Cardiac complications of arteriovenous fistulas in patients with end-stage renal disease. Nefrologia. **35**(3), 234–245 (2015)
163. Narechania, S., Tonelli, A.: Hemodynamic consequences of a surgical arteriovenous fistula. Ann Am Thorac Soc. **13**(2), 288–291 (2016)
164. Munch, P., Andresen, M., Brown, A.: Rapid resetting of aortic baroreceptors in vitro. Am J Phys. **244**(5), H672–H680 (1983)
165. Lohmeier, T., Lohmeier, J., Warren, S., May, P., Cunningham, J.: Sustained activation of the central baroreceptor pathway in angiotensin hypertension. Hypertension. **39**(2 Pt 2), 550–556 (2002)
166. Thrasher, T.: Effects of chronic baroreceptor unloading on blood pressure in the dog. Am J Physiol Regul Integr Comp Physiol. **288**(4), R863–R871 (2005)
167. Brooks, V., Sved, A.: Pressure to change? Re-evaluating the role of baroreceptors in the long-term control of arterial pressure. Am J Physiol Regul Integr Comp Physiol. **288**(4), R815–R818 (2005)
168. Lohmeier, T., Irwin, E., Rossing, M., Serdar, D., Kieval, R.: Prolonged activation of the baroreflex produces sustained hypotension. Hypertension. **43**(2), 306–311 (2004)
169. Barrett, C., Ramchandra, R., Guild, S., Lala, A., Budgett, D., Malpas, S.: What sets the long-term level of renal sympathetic nerve activity: a role for angiotensin II and baroreflexes? Circ Res. **92**(12), 1330–1336 (2003)
170. Andressen, M., Yang, M.: Arterial baroreceptor resetting: contributions of chronic and acute processes. Clin Exp Pharmacol Physiol. **16**(Suppl 15), 19–30 (1989)
171. Caria, S., Cupisti, A., Sau, G., Bolasco, P.: The incremental treatment of ESRD: a low-protein diet combined with weekly hemodialysis may be beneficial for selected patients. BMC Nephrol. **15**, 172 (2014)
172. Daugirdas, J., Blake, G., Ing, T. (eds.): Handbook of Dialysis, 5th edn. Wolters Kluwer Health, Philadelphia (2015)
173. Morel, H., Jaffrin, M., Lux, C., Renou, M., Fessier, C., Petit, A., Morinière, P., Legallais, C.: A comparison of bicarbonate kinetics and acid-base status in high flux hemodialysis and on-line post-dilution hemodiafiltration. Int J Artif Organs. **35**(4), 288–300 (2012)
174. Lagarias, J., Reeds, J., Wright, M., Wright, P.: Convergence properties of the Nelder-Mead simplex method in low dimensions. SIAM J Optim. **9**(1), 1112–1147 (1998)

Chapter 3
Mean Arterial Blood Pressure During Haemodialysis: Sensitivity Analysis and Validation of Mathematical Model Predictions

Abstract In this chapter we analyse and assess the proposed compartmental model of the cardiovascular system and body fluid shifts during haemodialysis treatment. The local sensitivity of the main model outcome, i.e. the mean arterial blood pressure, to changes in model parameters is studied in order to identify the key model parameters and their impact on simulation results. Model validation using clinical data is also presented along with a discussion on model limitations and its shortcomings.

Keywords Model parameters · Sensitivity analysis · Local sensitivity · Model validation · Data fitting · Model limitations · Pulsatility · Sympathetic-parasympathetic interaction · Baroreceptor resetting · Humoral regulation · Regional blood circulations · Pulmonary circulation · Autoregulation

3.1 Sensitivity Analysis

3.1.1 Overview

Due to the complexity of the model and the used modelling approach, the analysis of local model sensitivity to individual parameters is not straightforward.

Firstly, there are different groups of parameters to be considered. In the present model, the values of parameters were assigned or assumed based on physiological data from the literature. However, only some of these 'assigned parameters' are used directly in the model equations. In many cases the values of 'equation parameters' (i.e. the parameters seen in model equations) are calculated based on the values of one or more 'assigned parameters' (as described in Chap. 2) or are calculated numerically to ensure initial steady-state conditions of the modelled system (also described in Chap. 2).

For instance, Eq. (2.50) defines the hydrostatic pressure of the interstitial fluid (P_{is}) as a linear function of the interstitial volume (V_{is}):

L. Pstras, J. Waniewski, *Mathematical Modelling of Haemodialysis*,
https://doi.org/10.1007/978-3-030-21410-4_3

$$P_{is} = P_{is,n} + \frac{1}{C_{is}} \cdot (V_{is} - V_{is,n})$$

The above equation contains three parameters: $P_{is,n}$, the normal interstitial pressure; $V_{is,n}$, the normal interstitial volume; and C_{is}, the interstitial compliance. Only one of these parameters ($P_{is,n}$) is an 'assigned parameter', whereas the other two are 'calculated parameters', which depend on other (assigned or calculated) parameters as shown in Fig. 3.1.

Another example can be given for the calculation of the blood flow rate between compartments A and B ($Q_{A\text{-}B}$) based on Eqs. (2.2), (2.4), and (2.9):

$$Q_{A\text{-}B} = \frac{P_A - P_B}{R_A}$$

$$P_A = \frac{(V_A - V_{u,A})}{C_A}, \quad P_B = \frac{(V_B - V_{u,B})}{C_B}, \quad R_A = \frac{\kappa_A(1 + 2.5 \cdot HCT_A)}{V_A^2}$$

The above set of equations contains five parameters: $V_{u,A}$ and $V_{u,B}$, unstressed volumes of compartments A and B; C_A and C_B, compliances of compartments A and B; and κ_A, parameter related to the hydraulic resistance of compartment A (the remaining quantities, i.e. V_A, V_B and HCT_A, are state variables). The values of all these parameters are not explicitly assigned but are calculated based on several different parameters, as shown in Fig. 3.2.

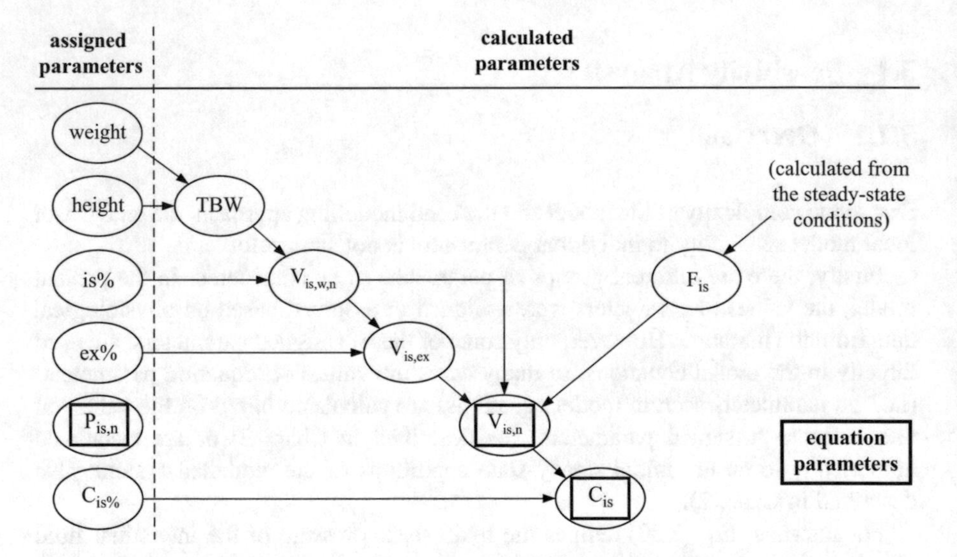

Fig. 3.1 Schematic representation of the process of assigning model parameters related to the interstitial hydrostatic pressure. TBW, total body water; $V_{is,w,n}$, normal interstitial water volume; is %, interstitial fraction of TBW; $V_{is,ex}$, interstitial volume excluded to proteins; ex%, $V_{is,ex}$ as percentage of $V_{is,w,n}$;; F_{is}, water fraction of the interstitial fluid available to proteins; $V_{is,n}$, normal interstitial volume; C_{is}, interstitial compliance; $C_{is\%}$, interstitial compliance as % of $V_{is,n}$

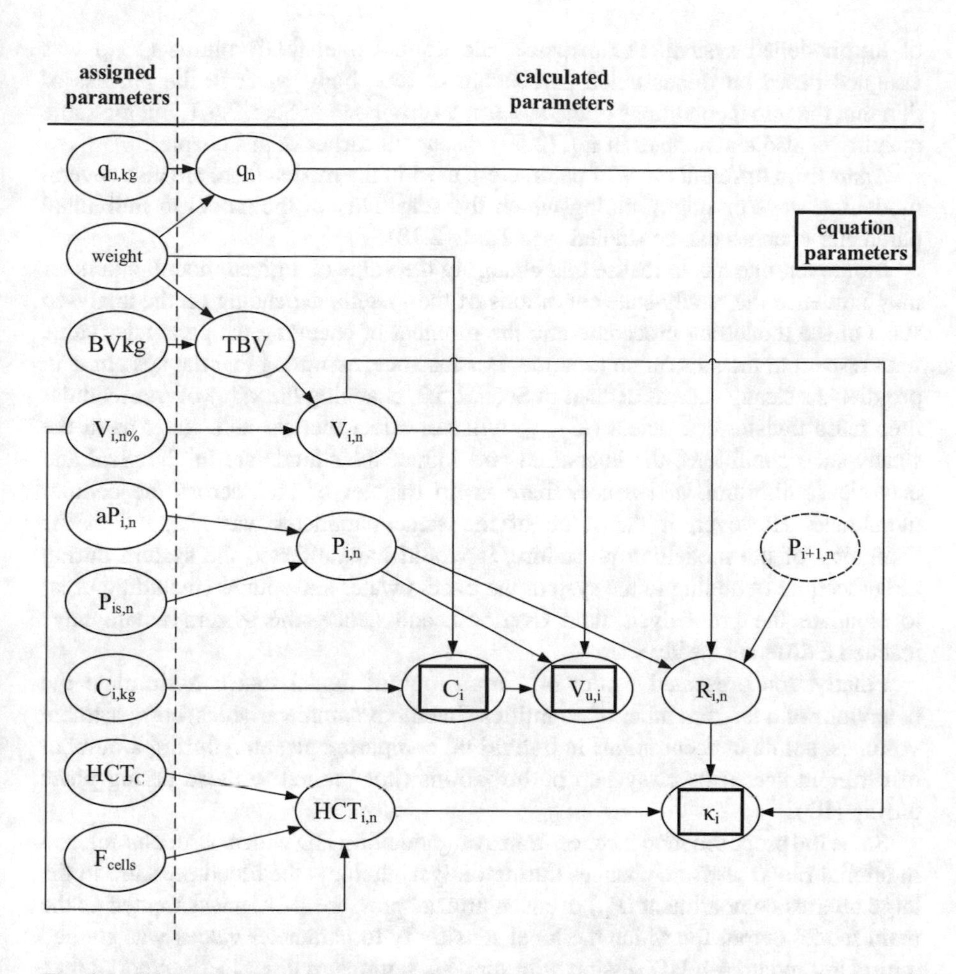

Fig. 3.2 Schematic representation of the process of assigning model parameters related to the normal vascular pressure and resistance of the *i*-th blood compartment. q_n, normal cardiac output; BV_{kg}, blood volume per kg of body weight; TBV, total blood volume; *V*, vascular volume; *P*, transmural blood pressure; aP, absolute blood pressure; P_{is}, interstitial pressure; *C*, compliance; V_u, unstressed volume; *R*, vascular resistance; HCT_C, central haematocrit; F_{cells}, ratio of whole-body to central haematocrit; κ, parameter related to the vascular resistance

Similarly, the protein permeability-surface product of the capillary wall, i.e. the parameter PS_p (PS_{alb} or PS_{glob}) in Eq. (2.62), is not assigned explicitly but is calculated from the steady-state conditions based on the assigned values of transcapillary escape rate of albumin and globulins (TER_{alb} and TER_{glob}) and other parameters determining protein leakage through the capillary walls (i.e. protein reflection coefficients, normal plasma protein concentration, normal plasma volume, normal lymph flow).

Some of the parameters, such as normal volumes or pressures of individual compartments, can be treated as both parameters of the model and initial conditions of the modelled system. For instance, the normal interstitial volume ($V_{is.n}$) was assigned based on the assumed percentage of total body water in the process of defining the initial conditions of the system, as discussed in Sect. 2.5.1, but the same quantity is also a parameter in Eq. (2.50) discussed earlier in this chapter.

Apart from different types of parameters used in the model, there are also several modelling steps or stages, during which the sensitivity of the model to individual parameter changes can be studied (see Table 2.12).

Moreover, one has to realise that changing the value of a given model parameter may influence the steady-state conditions of the system, depending on the analysed stage of the modelling procedure and the moment of changing the parameter value with respect to the simulation timeline. For instance, assuming that the system is in pre-dialysis steady state as defined in Sect. 2.5.2, changing the value of transcellular urea mass transfer coefficient ($K_{U,cell}$) will not affect the system's state; as in the steady-state conditions, the interstitial and intracellular fluids are in chemical and osmotic equilibrium, and hence, there is no transfer of urea across the cellular membrane. However, if the value of the same parameter was changed at the beginning of the modelling procedure, it would have affected the system during the procedure of adding to the system the excess water and solutes (including urea) to simulate the pre-dialysis fluid overload, and hence, the system would have reached a different steady state.

Finally, the presented model is a multi-output model that can simulate the behaviour of a large number of quantities (haemodynamic variables, compartment volumes, solute concentrations in individual compartments, etc.) during a number of different scenarios or system perturbations (not limited to those taking place during HD).

Since the proposed model concentrates on modelling HD with a particular interest in arterial blood pressure changes during dialysis therapy, the blood pressure in the large arteries compartment (P_{la}) or mean arterial pressure (MAP) was treated as the main model output for which the local sensitivity to parameter values was studied during a standard 4-h HD session with dialysis settings as described in Sect. 2.6.

Given the aforementioned differences between the types of parameters in the model, two separate local sensitivity analyses were performed: (1) analysis of MAP sensitivity to 'assigned parameters' and (2) analysis of MAP sensitivity to 'equation parameters'.

The parameters related strictly to HD (i.e. solute dialysances or clearances, composition of the dialysate fluid, priming volume of the extracorporeal circuit, dialyzer blood flow rate, ultrafiltration rate and the timing of the dialysis procedure) were treated not as model parameters but as HD settings or characteristics and hence were not included in the sensitivity analysis. A separate study was performed to show the impact of different dialysis settings on the modelled cardiovascular system (see Sect. 4.7).

3.1.2 Sensitivity to Assigned Parameters

The dimensionless relative local sensitivity of the output y_k (in this case MAP) to the i-th parameter (θ_i) was calculated as follows:

$$S_{i,k} = \frac{\partial y_k}{\partial \theta_i} \cdot \frac{\theta_i}{y_k}\bigg|_{\theta=\theta_0}, \quad \theta_i, y_k \neq 0 \tag{3.1}$$

The derivative in the above equation was computed using the central difference approximation.

For each studied parameter θ_i, the following procedure was applied. First, the given parameter was increased by 0.01% with all other parameters unchanged, and the whole modelling procedure (all seven steps, as described in Table 2.12) was repeated from the beginning. Subsequently, the same parameter was decreased from the original value by the same percentage (again with all other parameters unchanged), and the whole modelling procedure was repeated once again. The resulting MAP profiles during the last modelling stage (i.e. during the HD procedure including filling of the extracorporeal circuit with the patient's blood) from both simulations along with the original MAP profile (from the dialysis simulation without any parameter changes) were then used to calculate the quasi-continuous dimensionless relative local sensitivity of MAP to the given parameter throughout the dialysis procedure using Eq. (3.1). This approach does not only identify the key model parameters but also shows how their impact changes over the time of the simulated procedure.

Sensitivity to the following 'assigned parameters' was evaluated:

- General parameters (weight, normal cardiac output, normal heart rate, mean intrathoracic pressure)
- Parameters describing water distribution in the body (total body water, fractions of total body water assigned to interstitial fluid, intracellular fluid and blood, intracellular fluid water fraction)
- Parameters of the blood compartments (normal blood volumes, normal absolute pressures, mass specific compliances, normal plasma water fraction, normal RBC water fraction)
- Red blood cells parameters (normal central haematocrit, ratio of whole-body to central haematocrit, mean corpuscular volume of RBCs, thickness of RBCs)
- Parameters of the left and right cardiac stroke volume curves (maximal SV, sensitivity of SV to atrial pressure, parameter m describing the impact of afterload)
- Parameters of all baroreflex mechanisms (gains, amplitudes, time constants)
- Parameters of extravascular fluid compartments and lymphatic system (normal interstitial hydrostatic pressure, interstitial compliance, percentage of the interstitial space excluded to proteins, normal lymph flow, lymph flow sensitivity to interstitial pressure changes)

- Transport parameters of the capillary wall (total capillary surface area, whole-body hydraulic conductivity, solute permeabilities, solute reflection coefficients, transcapillary escape rates of proteins)
- Transport parameters of the tissue cell membrane and RBC membrane (solute and water mass transfer coefficients or permeabilities, reflection coefficients of small solutes, solute equilibrium concentration ratios)
- Solute charges and osmotic coefficients
- Initial plasma solute concentrations and protein concentrations in tissue cells and red blood cells
- Molecular weights and density of proteins
- Urea and creatinine generation rates
- Gibbs-Donnan coefficients for the dialyzer membrane

One hundred thirty-nine parameters were studied in total. The results of the local sensitivity analysis are shown in Fig. 3.3. The figure shows only the parameters to which MAP is the most sensitive during the dialysis procedure, i.e. the parameters for which the difference between the relative sensitivity at time 0 and relative sensitivity at the end of the simulation is greater than 0.05 (other parameters with relatively small, near-zero sensitivities were omitted for clarity). These 'important' or 'critical' parameters constituted slightly above 10% of all studied parameters (16 out of 139).

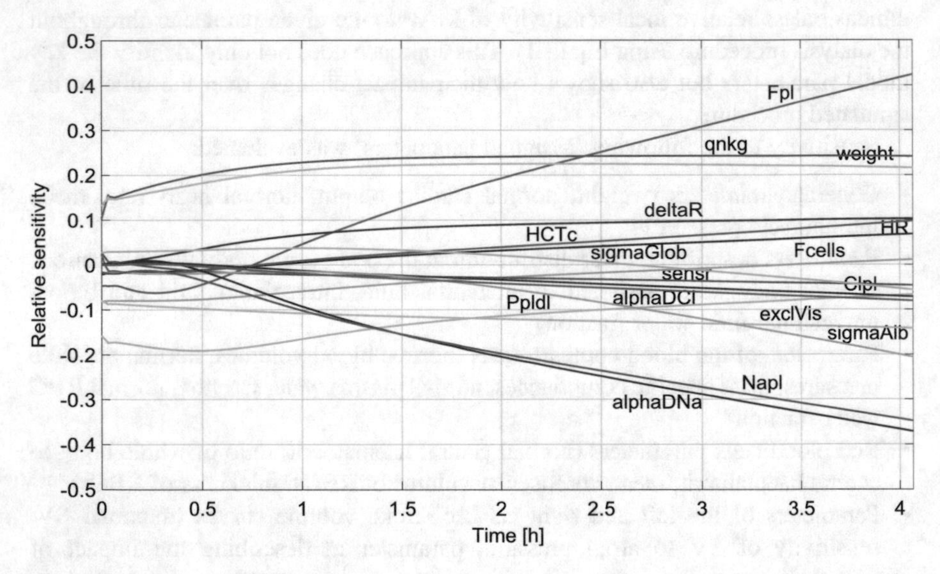

Fig. 3.3 Relative sensitivity of mean arterial pressure to assigned model parameters at the pre-dialysis steady-state conditions (time 0) and during a 4-h haemodialysis procedure including filling of the extracorporeal circuit with blood. Symbols: F_{pl}, plasma water fraction; $q_{n,kg}$, normal cardiac output per kg of body weight; ΔR, amplitude of arterioles resistance; HCT_C, normal central haematocrit; HR, normal heart rate; F_{cells}, ratio of whole-body haematocrit to central haematocrit; $sens_r$, sensitivity of right ventricular stroke volume to right atrial pressure; $\sigma_{Alb}/\sigma_{Glob}$, capillary wall reflection coefficient of albumin/globulins; TPP, total plasma protein; $exclV_{is}$, percentage of interstitial space excluded to proteins; Cl_{pl}, plasma chloride concentration; Na_{pl}, plasma sodium concentration; $\alpha_{DNa}/\alpha_{DCl}$, Gibbs-Donnan coefficient for sodium at the dialyzer membrane

At time 0 one can see the sensitivity of MAP to the studied parameters at pre-dialysis steady-state conditions. At this point, for most of the parameters, the absolute values of relative sensitivities are below 0.1 meaning that a change in the parameter value by 10% will affect the studied model outcome by less than 1%. For all studied parameters (including those not shown in Fig. 3.3), the sensitivity of MAP at time 0 was lower than 0.2, indicating that despite the model complexity and the high number of parameters, the model is relatively robust and resistant to changes of parameter values.

Beyond time 0, Fig. 3.3 shows how the sensitivity of MAP to changed parameter values varies throughout the whole dialysis procedure including filling of the extracorporeal circuit with the patient's blood. At the end of the 4-h dialysis, the absolute values of relative sensitivity to all parameters are still below 1, meaning that a change in any parameter value does not affect the model outcome by more than the actual parameter change (in percentage terms).

As shown in Fig. 3.3, the three parameters with the highest local influence on MAP during dialysis are initial concentration of sodium in blood plasma ($Na_{pl} =$ 140 mmol/L), initial plasma water fraction ($F_{pl} = 0.94$) and the Gibbs-Donnan coefficient for sodium at the dialyzer membrane ($\alpha_{DNa} = 0.94$). The three above parameters determine the concentration of sodium in plasma water available for transport (i.e. not attracted to proteins) and hence determine the amount of sodium exchanged between plasma and dialysate fluid according to Eqs. (2.76) and (2.77) (additionally, F_{pl} influences also the total blood volume, which results in a higher sensitivity of MAP to this parameter). The higher the Na_{pl} or the lower the F_{pl}, the higher the sodium concentration in plasma water, whereas the higher the α_{DNa}, the more sodium is freely available for transport. Since the concentration of sodium in the dialysate fluid was fixed ($Na_d = 142$ mmol/L), the higher the effective concentration of sodium in plasma water, the more sodium is lost from plasma to dialysate, which leads to the reduction of plasma osmolarity and a reduced plasma refilling rate, causing eventually a higher reduction in plasma volume and blood pressure. The importance of sodium balance for the cardiovascular stability is shown also in Sect. 4.7 through varying the concentration of sodium in the dialysate fluid.

The above applies also to the exchange of chloride between plasma and dialysate fluid, but since the concentration of chloride in plasma is typically reduced during dialysis (as opposed to the relatively stable concentration of sodium) [1], the initial plasma concentration of chloride ($Cl_{pl} = 103$ mmol/L) has a lower negative impact on MAP, compared to Na_{pl} (again, assuming a fixed chloride concentration in the dialysate fluid $Cl_d = 108$ mmol/L).

From Fig. 3.3 it seems that F_{pl} is the most influential parameter for the mean arterial pressure during haemodialysis. Please note, however, that the relative sensitivity of MAP to this parameter is only 0.5, meaning that a change of 1% in F_{pl} (say, from 0.94 to 0.93) would induce only a 0.5% change in MAP (a 0.5% decrease in this case). Moreover, in reality the value of F_{pl} should not vary by more than 1–2% (depending on the amount of proteins and other solids in plasma), and hence, even if the model seems to be sensitive to this parameter, its relatively low potential changes reduce its relative importance. The above applies also to other parameters, which are not expected to differ largely from the values assumed in the model (e.g. the Gibbs-Donnan coefficients).

Three of the parameters shown in Fig. 3.3 are related to cardiac output – the normal cardiac output per kg of body weight ($q_{n,kg} = 75$ mL/min/kg), the normal heart rate (HR = 75 bpm) and the sensitivity of right ventricular stroke volume to changes in right atrial pressure ($sens_r = 12$ mL/mm Hg). The latter affects the response of the heart to the reduction of cardiac preload during dialysis. The more sensitive the heart to changes in atrial pressure, the higher the reduction of stroke volume during dialysis and hence a higher decrease in the mean arterial pressure. Given that the sensitivity of right ventricular stroke volume to atrial pressure changes is normally higher than that of the left ventricle [2], it is not surprising that the model is more sensitive to this parameter than to that of the left ventricle. The normal cardiac output determines normal vascular resistances (based on Eq. (2.2) and assumed vascular pressures). The higher the cardiac output, the higher the initial (normal) resistance of small arteries and arterioles and hence the bigger the impact of baroreflex regulation of systemic arteries in response to blood volume reduction. Analogically, the higher the normal heart rate, the lower the normal heart period and hence the bigger the impact of baroreflex regulation of heart period during HD. Among the baroreflex parameters, MAP is the most sensitive to the amplitude of systemic resistance regulation ($\Delta R = 1.0$ mmHg∗s/mL).

The patient's weight (assumed 70 kg) can be treated as the scaling factor in the model, which influences the total body water (and hence the volumes of all compartments), the cardiac output and the compliances of all vascular compartments, and therefore, it is at no surprise that this parameter has some, albeit relatively small, impact on the studied model outcome.

The remaining of the parameters identified by the sensitivity analysis, i.e. the initial concentration of total plasma proteins (TPP = 7 g/dL), the capillary wall reflection coefficient for albumin ($\sigma_{Alb} = 0.95$) and for globulins ($\sigma_{Glob} = 0.98$), the normal central haematocrit ($HCT_C = 0.47$), the ratio of whole-body haematocrit to central haematocrit ($F_{cells} = 0.9$) and the percentage of the interstitial fluid volume excluded to proteins ($exclV_{is} = 50\%$) influence MAP mainly through their impact on the transcapillary protein transport and the vascular refilling mechanism. Since the initial transcapillary rates of escape of albumin and globulins (TER_{Alb} and TER_{Glob}) are expressed as percentage of the amount of albumin and globulins in the total plasma volume, the higher the TPP, the more proteins leak through the capillary walls and therefore the higher their concentration in the interstitial fluid, which affects the interstitial oncotic pressure and the transcapillary water filtration/absorption processes. Similarly the higher the initial total plasma volume, the more proteins are contained within the vascular space and hence the higher their transcapillary escape rate. HCT_C and F_{cells} influence the initial total plasma volume – the higher the HCT_C or the F_{cells} ratio, the higher the red blood cell volume and hence the higher the plasma volume (given the assumed amount of water contained in the blood and the assumed plasma and RBC water fractions, with a higher volume of RBC, more blood water must be contained in the plasma). The albumin reflection coefficient (σ_{Alb}) determines the relationship between the convective and diffusive transcapillary leakage of albumin. At the given TER_{Alb}, the higher the σ_{Alb}, the more proteins leak diffusively, and hence, when during dialysis the transcapillary filtration reverses into absorption of fluid from the interstitium to plasma, less proteins are absorbed convectively from the interstitium, whereas more proteins continue to leak diffusively, leading overall to a lower plasma protein concentration and a lower vascular refilling rate. In general, the model is more sensitive to parameters related to albumin, compared to globulins, given that albumin is more abundant and generates a higher oncotic pressure. Finally, with a higher percentage of the normal interstitial fluid excluded to proteins ($exclV_{is}$), the volume of the interstitial space containing proteins is accordingly lower and subject to relatively higher volume reduction due to dialyzer ultrafiltration, which increases to a larger extent the interstitial oncotic pressure and hence reduces the transcapillary absorption of the interstitial fluid.

The above remarks on the influence of individual parameters on MAP during dialysis are the most evident or plausible explanations of the results of this sensitivity analysis; however, given the complexity of the proposed model and the number of interactions between different mechanisms and model features, there may be a more complex explanation of the magnitude of impact of individual assigned parameters on MAP.

Note also that the analysis presented in this chapter (as well as in the next chapter) studies the local sensitivity of the given model output to small parameter changes around the assumed normal values. A similar analysis could be performed for different sets of parameter values (e.g. using Monte Carlo techniques). Moreover, apart from the local sensitivity analysis, a global sensitivity analysis could be performed to investigate the impact of large parameter changes, especially for those parameters which may feature a large variation (please see Sect. 4.6 for the analysis of the impact of large changes in baroreflex and cardiac parameters).

3.1.3 Sensitivity to Equation Parameters

The local sensitivity of intradialytic MAP to changes in 'equation parameters' was studied using the similar dimensionless local sensitivity approach presented in the previous chapter. The analysis was started with the system in the pre-dialysis steady-state conditions, as described in Sect. 2.5.2. For each studied parameter θ_i, the following procedure was then applied. Firstly, the given parameter was increased by 0.01% with all other parameters unchanged, the system was simulated for 192 h to reach new steady-state conditions, and the whole HD procedure was simulated as described before. Subsequently, the same parameter was decreased from the original value by the same percentage (again with all other parameters unchanged), and again the new steady-state condition of the system was found, and the HD procedure was simulated. The same procedure, i.e. a 192-h simulation to reach the steady-state conditions and the standard simulation of the HD session, was performed for the system with the original value of parameter θ_i. The resulting MAP profiles during the last modelling stage (i.e. during the whole HD procedure including filling of the extracorporeal circuit with blood) from all three simulations were then used to calculate the quasi-continuous dimensionless relative sensitivity of MAP to the given parameter throughout the dialysis procedure using Eq. (3.1).

Sensitivity to the following 'equation parameters' was evaluated:

- Unstressed volumes ($V_{u,i}$) and compliances (C_i) of all cardiovascular compartments [Eq. (2.4)]
- Parameters κ_i describing the vascular resistances [Eq. (2.9)]
- Initial tube haematocrit in all cardiovascular compartments ($HCT_{T,j,0}$) [Eq. (2.31)]
- Parameters of the left and right cardiac stroke volume curves (SV_{max}, x_r, x_l, s_r, s_l, m, $P_{la,n}$, $P_{pa,n}$) [Eqs. (2.11) and (2.12)]
- Parameters of all baroreflex mechanisms (gains, maximal and minimal levels of controlled variables, time constants, slopes of the sigmoidal relationships) [Eqs. (2.13)–(2.27)]
- Parameters of the interstitial fluid and lymphatic system ($V_{is,n}$, $P_{is,n}$, C_{is}, $Q_{L,n}$, LS, $P_{is,ex}$, $V_{is,ex}$) [Eqs. (2.50), (2.63), and (2.64)]
- Transport parameters of the capillary wall ($p_{s,cap}$, S_{cap}, σ_s, Lp) [Eqs. (2.38) and (2.48)]
- Transport parameters of the tissue cell membrane ($K_{s,cell}$, $\beta_{s,cell}$, $\sigma_{s,cell}$, $K_{w,cell}$) [Eqs. (2.33), (2.34), and (2.65)]
- Transport parameters of the RBC membrane ($p_{s,rc}$, $\beta_{s,rc}$, $\sigma_{s,rc}$, d, $K_{w,rc}$) [Eqs. (2.41), (2.43), and (2.71)]
- Solute charges (z_s) [Eq. (2.37) and similar for other membranes]
- Solute osmotic coefficients (φ_s, $\varphi_{p,ic}$) [Eqs. (2.48), (2.66), (2.67), (2.72), and (2.73)]
- Molecular weights and density of albumin and globulins (MW_{Alb}, MW_{Glob}, ρ_{Alb}, ρ_{Glob}) [Eq. (2.70) and conversions between molar and mass concentrations of proteins]
- Urea and creatinine generation rates (g_u, g_{Cr})
- Gibbs-Donnan coefficient for the dialyzer membrane (σ_d) [Eqs. (2.76) and (2.77)]

The sensitivity of MAP to the transcapillary Gibbs-Donnan ratio for monovalent cations (α_{cap}) was not studied given that this parameter is computed from the steady-state conditions for the healthy subject, and hence, it cannot be treated as an independent parameter.

One hundred twenty-three parameters were studied in total. The results of the sensitivity analysis are shown in Fig. 3.4. Again, the figure shows only the parameters to which MAP is the most sensitive during the dialysis procedure, i.e. the parameters for which the difference between the relative sensitivity at time 0 and relative sensitivity at the end of the simulation is greater than 0.1 (other parameters with relatively small sensitivities were omitted for clarity). These 'important' or 'critical' parameters constituted also slightly above 10% of all studied parameters (13 out of 123).

Similarly to the previous case, at time 0 one can see the relative sensitivity of MAP to the studied parameters at pre-dialysis steady-state conditions. At this point,

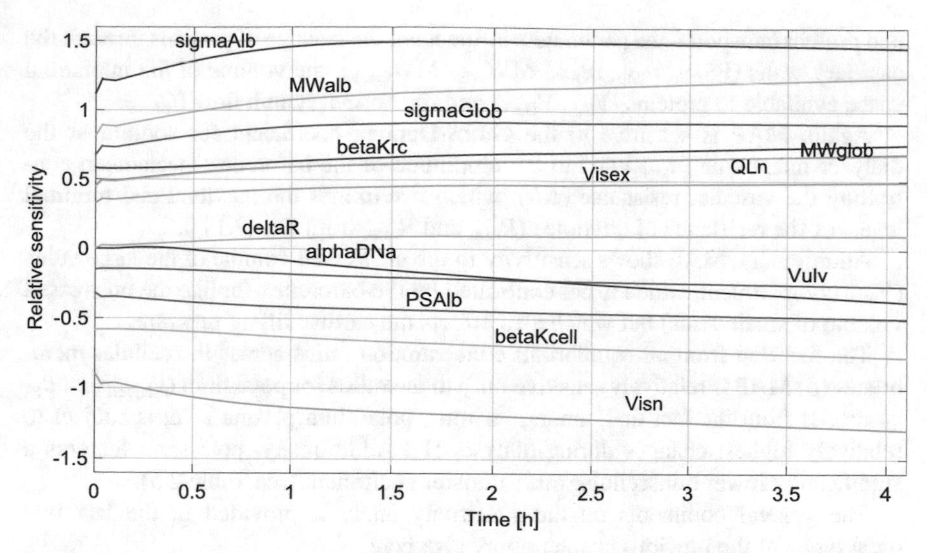

Fig. 3.4 Relative sensitivity of mean arterial pressure to parameters of model equations at the pre-dialysis steady-state conditions (time 0) and during a 4-h haemodialysis procedure including filling of the extracorporeal circuit with blood. Symbols: $\sigma_{Alb}/\sigma_{Glob}$, capillary wall reflection coefficient of albumin/globulins; MW_{Alb}/MW_{Glob}, molecular weight of albumin/globulins; $\beta_{K,rc}/\beta_{K,cell}$, equilibrium concentration ratio of potassium across the tissue cell/RBC membrane; QL_n, normal lymph flow; Vis_{ex}, the volume of interstitial fluid excluded to proteins; ΔR, amplitude of arterioles resistance; α_{DNa}, Gibbs-Donnan coefficient for sodium at the dialyzer membrane; $V_{u,lv}$, unstressed volume of large veins; PS_{Alb}, albumin permeability-surface product of the capillary walls; $V_{is,n}$, normal volume of the interstitial fluid

the absolute values of MAP sensitivity to the studied parameters are in most cases slightly higher than in the previous analysis, which can be explained by the fact that the 'equation parameters' affect the modelled system to a higher extent, especially during the simulation of fluid overloading of the patient. Also, when the value of an equation parameter is changed, none of other parameters are changed, whereas changing one of the assigned parameters may influence other dependent parameters, which can have an opposite effect on the MAP. However, only a few of the relative sensitivities to equation parameters at time 0 are above 1, whereas the sensitivity to the vast majority of the studied parameters (including the parameters not shown in Fig. 3.4) is below 0.2, again indicating the relative robustness of the model.

Beyond time 0, Fig. 3.4 shows how the sensitivity of MAP to the equation parameters varies throughout the whole dialysis procedure including filling of the extracorporeal circuit with blood. At the end of a 4-h dialysis, only for a few of the parameters, the absolute values of the relative sensitivity are above 1, whereas for most of the parameters (including those not shown), the values are below 1, meaning again that a change in the parameter value does not affect the model outcome by more than the actual parameter change (in percentage terms).

As in the analysis of MAP sensitivity to 'assigned parameters', most of the 'equation parameters' that MAP is sensitive to are again related to vascular refilling and protein transport – the parameters influencing the leakage of proteins through the capillary walls (PS_{Alb}, σ_{Alb}, σ_{Glob}, MW_{Alb}, MW_{Glob}), the volume of the interstitial space available to proteins ($V_{is,n}$, $V_{is,ex}$) and the normal lymph flow ($Q_{L,n}$).

Again, MAP is sensitive to the Gibbs-Donnan coefficient for sodium at the dialyzer membrane (α_{DNa}) and to the amplitude of the baroreflex mechanism controlling the vascular resistance (ΔR), which determines the maximal and minimal levels of the resistance of arterioles (R_{min} and R_{max} from Eq. (2.14)).

Additionally, MAP shows sensitivity to the unstressed volume of the large veins (V_{ulv}) compartment, which is not controlled by the baroreflex (unlike the unstressed volume of small veins) but which also affects the cardiac filling pressure.

The fact that from all equilibrium concentration ratios across the cellular membranes (β) MAP is relatively sensitive only to the ratios for potassium ($\beta_{K,cell}$ and $\beta_{K,rc}$) results from the fact that, among all ions, potassium plasma level is subject to relatively highest changes during dialysis [1]. Additionally, potassium features a significantly lower transcellular mass transfer coefficient (see Table 2.5).

The general comments on the sensitivity analysis provided in the last two paragraphs of the previous chapter apply also here.

3.2 Model Validation

The model was validated through testing its ability to fit clinical data describing haemodynamics and solute kinetics during haemodialysis. The data used for model fitting came from (1) past measurements performed on a group of dialysed patients by the collaborators of our research group at Medical University of Lublin, Poland (funded by the Polish National Science Centre, grant No. N N518 289840) and (2) a past study published in the literature.

Please see Chap. 5 for a short discussion on how the model could be further validated in the future and on the challenges associated with collecting data from dialysis patients.

3.2.1 *Lublin Data*

3.2.1.1 Patients

The data presented below comes from 12 anuric patients (see Table 3.1) with arteriovenous fistula undergoing maintenance, thrice-weekly HD at the Lublin Medical University, Poland. The data was collected during three consecutive HD sessions of a 1-week dialysis treatment with the interdialytic breaks before the sessions of 3, 2 and 2 days, respectively. Written informed consent was obtained from each patient, and the study was approved by the Bioethical Committee of the Lublin Medical University. During all dialysis sessions, the patients remained in the

Table 3.1 General characteristics of the analysed group of 12 patients (mean ± SD, range)

Age	63.8 ± 11.3 (47–79)	years
Females/males	9/3	–
Time on dialysis	11.5 ± 10.2 (1–32)	years
Height	165 ± 8 (147–175)	cm
Weight before dialysis	72.6 ± 17.4 (45.4–103.2)	kg
Weight after dialysis	70.0 ± 16.9 (43.3–99.2)	kg

supine position. The studied group of patients has already been subject of several studies on intradialytic solute transport processes, such as phosphate kinetics [3, 4], phosphate, urea and creatinine clearances [5], extracellular calcium mass balance [6] and transcapillary transport of fluid and proteins during HD [7, 8].

3.2.1.2 Dialysis Sessions

The original dataset included data from three consecutive HD sessions in 25 patients (75 sessions in total), which for the purpose of this study was limited to 22 sessions, during which the patients did not receive any fluid or drug infusions and did not consume any drinks or food (to avoid any confounding influence on their fluid status), the dialyzer settings were not changed throughout the session and the data did not have any missing data points or artefacts. One, two or three sessions in the week were included in the study for each of the 12 analysed patients. The data from all patients and all 22 HD sessions was combined and averaged, thus describing the average dialysis in the average patient.

The analysed HD sessions lasted approximately 4 hours (237 ± 12 minutes) with the dialyzer blood flow rate in the range 200–350 mL/min (283 ± 44 mL/min), dialysate flow rate of 500 mL/min and dialyzer ultrafiltration in the range 0.8–4.0 L (see Table 3.2). The priming saline (0.9% NaCl) filling the extracorporeal circuit before each dialysis was in all cases infused to the patient, when the circuit was filled with the patient's blood.

For different patients the dialysis therapy was delivered using different dialyzers including Fresenius Helixone® high-flux dialyzers (FX 60, FX 80), Fresenius Polysulfone® low-flux dialyzers (F7HPS, F8HPS), Gambro Polyflux™ low-flux dialyzer (17L) and B.Braun Diacap® high-flux dialyzer (HI PS 20). Likewise, several different dialysis machines were used including Fresenius 4008b, Fresenius 4008s, Gambro AK 95 S and Nikkiso DBB-05.

3.2.1.3 Blood Tests

Before and during dialysis, several blood samples were collected for laboratory measurements. The first blood sample was collected before the patient was connected to the dialysis machine. The following blood samples were collected

Table 3.2 Parameters of the HD treatment in the analysed group of patients (mean ± SD, range)

Ultrafiltration	2.8 ± 0.9 (0.8–4.0)	L
Dialyzer blood flow rate	283 ± 44 (200–350)	mL/min
Dialysate sodium	142 ± 2 (138–146)	mmol/L
Dialysate potassium	2.5 ± 0.5 (2.0–3.4)	mmol/L
Bloodlines filling volume	165 ± 12 (146–172)[a]	mL
Dialyzer filling volume	97 ± 13 (74–121)[a]	mL

[a]Estimated based on data available from the manufacturer product datasheets

every hour during dialysis and at the end of the session from the arterial port in the fistula.

The haematological analysis of blood samples was performed using the automatic haematology analyser Advia® 2120 (Siemens Healthcare, Erlangen, Germany) based on the following measurement methods [9]. The red blood cell count (RBC) and mean corpuscular volume of red blood cells (MCV) are measured utilising the principles of flow cytometry with hydrodynamic focusing and laser light scattering, preceded by isovolumetric cell sphering with a special reagent containing sodium dodecyl sulphate and glutaraldehyde, thus eliminating the cell shape variability factor [9] (MCV is obtained as mean value of the red blood cell volume histogram). The haemoglobin concentration (HGB) in the blood (following erythrocyte lysis) is measured using the cyanide-free HGB method with colorimetric readings [9]. According to the technical specification of the Advia® 2120 system, the coefficient of variation for the measurements of RBC, MCV and HGB are 1.2%, 0.78% and 0.93%, respectively. The haematocrit (HCT) is calculated as the product of RBC and MCV in the analysed sample.

The chemical analysis of blood samples was done using the clinical chemistry analyser Advia® 1800 (Siemens Healthcare, Erlangen, Germany).

3.2.1.4 HD Simulation

Before simulating HD for the average patient representing the analysed group of patients, the modelled system was brought to the initial pre-dialysis steady-state conditions (using the procedure described in Sect. 2.5.2) matching the parameters of blood measured in the analysed patients before dialysis (see Table 3.3).

Moreover, a few model parameters were changed to better represent the analysed average patient, as follows. The weight measured after dialysis (see Table 3.1) was assumed to correspond to the patients' dry weight and was treated as the normal weight in the model. The normal total body water (TBW) was calculated for each patient from the Hume formula [10] (separate for males and females) with the average value used as TBW in the model. The fractions of TBW corresponding to extravascular extracellular water and extravascular intracellular water were taken for each patient from [11] (separate values for males and females), and again the average value was used in the model. The excess pre-dialysis water was assumed to

Table 3.3 Pre-dialysis blood parameters in the analysed group of patients (mean ± SD, range)

Plasma sodium	140.1 ± 2.3 (137.0–146.0)	mmol/L
Plasma potassium	5.8 ± 0.7 (4.6–7.0)	mmol/L
Plasma urea	24.6 ± 6.4 (15.7–40.4)	mmol/L
Plasma creatinine	0.83 ± 0.2 (0.46–1.26)	mmol/L
Plasma albumin	4.0 ± 0.3 (3.4–4.4)	g/dL
Plasma total protein	6.8 ± 0.3 (6.1–7.5)	g/dL
Blood haemoglobin	11.3 ± 0.7 (10.3–13.2)	g/dL
Haematocrit	34.9 ± 2.6 (30.8–39.8)	%
Mean corpuscular volume of RBCs	93.7 ± 3.8 (87.0–101.0)	fL
Red blood cell count	3.7 ± 0.2 (3.4–4.4)	$\times 10^6$ cells/μL
Mean arterial blood pressure	95 ± 13 (71–119)	mm Hg

correspond to the difference between the dry weight (weight after dialysis) and weight measured before dialysis (see Table 3.1).

The dialysis duration, blood flow rate, ultrafiltration and the concentration of sodium and potassium in the dialysate fluid were set according to the data shown in Table 3.2. Since the data on other ions considered in the model (i.e. chloride, bicarbonate and other cations) was not available, their pre-dialysis plasma concentrations and dialysate concentrations were left as for the standard dialysis in the reference patient (see Tables 2.9 and 2.11 in Chap. 2).

Given that the first blood sample was collected before the patients were connected to the dialysis machine, the complete dialysis procedure was simulated including filling of the extracorporeal circuit with the patient's blood and simultaneous infusion of the priming saline into the patient. The average filling volumes of the dialyzer and the dialysis machine bloodlines used in different patients (as reported in Table 3.2) were calculated based on data available from the manufacturer product datasheets.

Instead of assuming a certain blood flow rate for filling of the extracorporeal circuit, it was assumed that the circuit is filled with the patient's blood within 2 minutes. During this time the normal saline (154 mmol/L of sodium + 154 mmol/L of chloride) filling the extracorporeal circuit is infused to the patient. One minute was assumed for the 'idle' circulation of blood in the extracorporeal circuit before the start of HD. The total simulation time was therefore 240 minutes (2 + 1 + 237).

Note that the assumed ultrafiltration (2.8 L) approximately matches the sum of the measured patient weight difference (2.6 kg) and the estimated volume of saline infused to the patient at the beginning of the HD procedure (~260 mL), taking into account that during dialysis several blood samples were collected for laboratory measurements (5–10 mL each). The volume of the collected blood samples was, however, not accounted for in the model simulations, as being negligible.

3.2.1.5 Fitting Procedure

The following data was available for fitting: morphologic blood parameters (HCT, HGB, MCV, RBC), biochemical plasma composition (sodium, potassium, urea, creatinine, albumin, total protein) and mean arterial blood pressure. Additionally, the plasma globulins concentration was calculated as the difference between total protein and albumin (as assumed in the model). The biochemical parameters of plasma were measured before and every hour during dialysis (five measurements in total). The morphologic blood parameters were measured only before and at the end of dialysis (to minimise the amount of blood collected from the patient). The blood pressure was measured before dialysis, at 2 hours into dialysis session and at the end of dialysis.

The aim of the fitting procedure was to match the clinical changes in the above 12 variables with the changes simulated by the model. For the sake of the simultaneous fitting of the above variables, all data and model simulations were normalised by the initial values. In total 34 data points were hence available for fitting (4 data points for each plasma solute measured every hour, 1 data point for each morphologic blood parameter measured at the end of dialysis and 2 data points for arterial blood pressure in the middle and at the end of the dialysis session). The aim was to minimise the sum of squared deviations between the simulated variables and clinical data (for all 12 variables), for which the Matlab built-in function *fminsearch* was used.

Initially 60 parameters (assigned parameters) were allowed to vary during the fitting procedure, including 3 aforementioned parameters adjusted before the procedure to represent the analysed group of patients (i.e. total body water and extra- and intracellular extravascular fractions of total body water), 3 parameters related to the dialysis procedure (access blood flow rate, urea diffusive clearance and the Gibbs-Donnan coefficient for sodium at the dialyzer membrane) and 54 parameters describing the cardiovascular system, baroreflex and the whole-body water and solute transport (see Table 3.4). The latter included the parameters identified earlier by the model sensitivity analysis in Chap. 3.1, but since the sensitivity analysis concentrated purely on arterial blood pressure, other parameters were included in the fitting procedure, such as the parameters related to the transport of measured solutes (i.e. sodium, potassium, urea, creatinine, albumin and globulins) and parameters related to water transport and distribution (e.g. hydraulic conductivity of the capillary walls, water transfer coefficients of RBCs and tissue cells, water fraction of RBCs and tissue cells, interstitial compliance or lymph flow sensitivity to interstitial pressure). All remaining model parameters are either related to other solutes (for which no data was available) or to the cardiovascular system but without significant impact on the arterial blood pressure (as shown by the sensitivity analysis), and hence, they were all left at the level set for the reference healthy subject (see Sects. 2.4 and 2.5.1). All assumptions related to model parameters, such as that the permeability of the capillary wall to ions is equal to urea permeability or that the diffusive dialysance of sodium and potassium is equal to the diffusive clearance of urea, were also left as set in the original model.

In order to avoid unrealistic values of the fitted parameters (i.e. negative values of parameters or reflection coefficients greater than 1), as well as to keep the parameters within a reasonable (biologically explainable) range, a modified version of the Matlab *fminsearch* function was used – *fminsearchbnd* – which allows for the use of boundary constraints. However, none of the final values of the fitted parameters approached closely the applied constraints.

To improve the efficiency of the fitting procedure, some of the fitted parameters were first adjusted 'manually' to obtain a rough visual fit between the model simulations and clinical data and hence to provide a better starting point for the fitting procedure.

3.2.1.6 Fitting Results

As shown in Fig. 3.5, the proposed model can fit the clinical data relatively well. The obtained level of fit (the sum of squared deviations between the model simulations and normalised data) was 77 percentage points squared, which translates into an average relative error of circa 1.5% (% of the initial value of each variable) for each of the fitted 34 data points. Using the Matlab *fminsearch* function does not guarantee finding the global optimum, and hence a better fit could be possibly achieved using a more complex and advanced fitting procedure with other optimisation tools (e.g. particle swarm optimisation) or a combination of different tools, but the achieved level of accuracy was deemed satisfactory for the purpose of this work.

As can be seen in Fig. 3.5, the solute data is generally slightly better fitted by the model than the morphologic parameters of blood. This is due to the higher number of solute data points, which creates an obvious bias of the fitting procedure towards the solute data.

Given that the performed model simulation corresponds to the whole dialysis procedure, including the infusion of the priming saline into the patient, one can clearly see the dilution of blood by the saline. The diluting effect is best visible for plasma proteins, which are generally confined to the vascular space with a relatively small transcapillary leakage and lymphatic refilling.

Table 3.4 presents the values of all fitted parameters and their relative change compared to the original values used in the model for the reference healthy subject. The fitted parameters can be divided into three groups.

The first group of parameters (22 parameters listed in the Section (A) of Table 3.4) were estimated at the values very similar to the values assumed in the original model. All these parameters can be practically removed from the fitting procedure (and left at the original level) with a very low impact on the quality of the overall fit (average relative error increased to 1.6%). Note that even if a given parameter does not influence considerably the overall model fit to all analysed variables, it can have an influence on the fit to individual variables. For instance, the decrease of mean corpuscular volume of red blood cells (MCV) could be better fitted through adjustments in the parameters describing the transport of ions across the capillary wall and

Table 3.4 Model parameters tuned to fit haemodynamic and solute data in the analysed patients

Parameter	Symbol	Unit	Ref. level	Fitted value	Δ [%]
Section (A)					
Total body water	TBW	L	34.9[a]	36.1	3.4
Extravascular intracellular water	icTBW	%	50.6[b]	52.1	2.9
Plasma water fraction	F_{pl}	–	0.94	0.94	0.2
Tissue cells water fraction	F_{ic}	–	0.70	0.73	1.5
Total capillary wall surface	S_{cap}	m^2	600	628	4.7
Globulins capillary reflection coefficient	$\sigma_{glob,cap}$	–	0.98	0.98	1.3
Albumin molecular weight	MW_{Alb}	g/mol	69 000	69 420	0.6
Globulins molecular weight	MW_{Glob}	g/mol	170 000	175 110	3.0
Transcellular Na^+ mass transfer	$K_{Na,cell}$	L/min	0.15	0.15	4.9
Transcellular K^+ mass transfer	$K_{K,cell}$	L/min	0.01	0.01	1.0
Transcellular Na^+ equilibrium ratio	$\beta_{Na,cell}$	–	0.088	0.090	2.9
Transcellular K+ equilibrium ratio	$\beta_{K,cell}$	–	29.19	29.51	1.1
RBC Na+ equilibrium ratio	$\beta_{Na,rc}$	–	0.13	0.13	1.3
RBC K+ equilibrium ratio	$\beta_{K,rc}$	–	28.33	28.76	1.5
Ratio of whole-body HCT to central HCT	F_{cells}	–	0.90	0.92	2.9
Lymph flow sensitivity to interstitial pressure	LS	mL/mmHg/h	43.0	45.1	4.9
Interstitial compliance	C_{is}	% of $V_{is,n}$/mmHg	12.0	12.5	3.8
% of P_{sv} transmitted to capillaries	w_v	%	80.0	82.5	3.1
Heart period amplitude	ΔT	s	0.70	0.69	−1.9
Heart contractility amplitude	ΔE	–	0.40	0.41	1.1
Dialyzer Gibbs-Donnan coefficient for sodium	α_{DNa}	–	0.94	0.94	0.2
Urea diffusive clearance	D_U	mL/min	210	211	0.3
Section (B)					
RBC water fraction	F_{rc}	–	0.72	0.66	−8.3
RBC water transfer coefficient	$K_{w,rc}$	L^2/min/mmol/m^2	7.6×10^{-5}	8.2×10^{-5}	8.4
Transcellular water transfer coefficient	$K_{w,cell}$	L^2/min/mmol	0.25	0.26	6.0
Capillary hydraulic conductivity	Lp	mL/min/mmHg	4.50	4.25	−5.5
Interstitial volume excluded to proteins	$V_{is,ex,\%}$	%	50.0	52.8	5.5
Small solute capillary reflection coefficient	$\sigma_{s,cap}$	–	0.05	0.06	21.0
Transcellular urea reflection coefficient	$\sigma_{U,cell}$	–	0.70	0.77	10.4
Transcellular creatinine reflection coefficient	$\sigma_{Cr,cell}$	–	0.70	0.81	15.7
RBC urea reflection coefficient	$\sigma_{U,rc}$	–	0.70	0.76	8.5

(continued)

Table 3.4 (continued)

Parameter	Symbol	Unit	Ref. level	Fitted value	Δ [%]
RBC creatinine reflection coefficient	$\sigma_{Cr,rc}$	–	0.70	0.76	8.9
RBC Na$^+$ permeability	$p_{Na,rc}$	cm/s	4.0×10^{-11}	5.0×10^{-11}	24.0
RBC K$^+$ permeability	$p_{K,rc}$	cm/s	2.0×10^{10}	3.5×10^{-10}	75.0
RBC urea permeability	$p_{U,rc}$	cm/s	3.0×10^{-4}	3.5×10^{-4}	16.7
RBC creatinine permeability	$p_{Cr,rc}$	cm/s	1.7×10^{-8}	2.2×10^{-8}	29.4
Urea capillary permeability	P_U	cm/s	2.2×10^{-5}	3.9×10^{-5}	77.3
Creatinine capillary permeability	P_{Cr}	cm/s	1.5×10^{-5}	1.8×10^{-5}	16.7
Urea generation rate	g_U	mmol/day	310	199	−35.9
Creatinine generation rate	g_{Cr}	mmol/day	9.8	11.3	15.3
Cardiopulmonary gain of *R* regulation	$G_{c,R}$	s/mL	0.63	0.77	22.2
Arterial gain of *R* regulation	$G_{a,R}$	s/mL	0.03	0.05	66.7
Arterial gain of $V_{u,sv}$ regulation	$G_{a,V}$	mL/mmHg	14.0	13.2	−5.6
Cardiopulmonary gain of *E* regulation	$G_{c,E}$	1/mmHg	0.10	0.14	39.5
Arterial gain of *E* regulation	$G_{a,E}$	1/mmHg	0.01	0.005	−46.0
Arterial sympathetic gain of *T* regulation	$G_{a,Ts}$	s/mmHg	0.01	0.012	17.0
Section (C)					
Extravascular extracellular water	exTBW	%	38.6 ‡	36.3	−6.0
Normal lymph flow	QL_n	L/day	8.0	7.7	−3.2
Normal cardiac output	q_{kg}	mL/min/kg	75.0	73.0	−2.7
Sensitivity of right SV to right atrial pressure	$sens_r$	mL/mmHg	12.0	17.1	42.6
Sensitivity of left SV to left atrial pressure	$sens_l$	mL/mmHg	6.0	10.4	74.0
Transcapillary escape rate of albumin	TER_{Alb}	%/h	5.0	5.7	14.8
Transcapillary escape rate of globulins	TER_{Glob}	%/h	3.0	4.4	48.1
Albumin capillary reflection coefficient	$\sigma_{alb,cap}$	–	0.95	0.99	3.8
Transcellular urea mass transfer	$K_{U,cell}$	L/min	0.80	0.20	−75.0
Transcellular creatinine mass transfer	$K_{Cr,cell}$	L/min	0.60	0.18	−70.7
Arterioles resistance amplitude	ΔR	mmHg∗s/mL	1.0	0.57	−43.0
Venous unstressed volume amplitude	ΔV	mL	600	805	34.2
Arterial vagal gain of *T* regulation	$G_{a,Tv}$	s/mmHg	0.02	0.01	−26.0
Cardiopulmonary gain of $V_{u,sv}$ regulation	$G_{c,V}$	mL/mmHg	208	3.1	−98.5

[a]Calculated from the Hume formula
[b]Estimated from Ref. [11]

RBC membrane, but this would not increase (or may even decrease) the global fit to all analysed variables, on which this fitting procedure was focused.

The second group of parameters (24 parameters listed in the Section (B) of Table 3.4) were estimated at values significantly different from the original values (although within biologically explainable and realistic limits); however, they had a relatively low impact on the overall quality of fit to data and could be also ignored in the fitting procedure.

Finally, 14 parameters shown in the Section (C) of Table 3.4 had a substantial impact on the overall quality of fit to the data, even if some of the estimated parameter values did not differ much from the original level (e.g. a small increase of albumin reflection coefficient of the capillary wall or a small decrease of the normal lymph flow). Within this group of parameters, the highest changes could be observed for the sensitivity of stroke volume to atrial pressure changes (an increase for both right and left heart), the transcapillary rates of escape of proteins (an increase for both albumin and globulins), the transcellular transfer coefficient for small solutes (a strong decrease for both urea and creatinine) and the baroreflex parameters, in particular a decrease in the amplitude of arterioles resistance (one of the parameters identified as important by the sensitivity analysis in Sect. 3.1.2), an increase in the amplitude of the venous unstressed volume and a strong decrease of the cardiopulmonary gain of the baroreflex mechanisms controlling venous unstressed volume. The latter is not surprising, as in one of the earlier models by Magosso et al. [12], the cardiopulmonary weight of the mechanism controlling venous capacity was set to zero.

Following the identification of the parameters substantially affecting the level of fit between model simulations and the analysed clinical data (14 parameters shown above), the fitting procedure was repeated with only these parameters being adjusted and all other model parameters set at the original level. The results of the second parameter fitting procedure are shown in Table 3.5.

Adjusting the values of only the above 14 parameters resulted in a similarly good level of fit of model simulations to the data, with the sum of squared deviations equal to 82 percentage points squared and the average error below 1.6% (the level of fit was visually indistinguishable; hence, no separate figure is shown). The values of the adjusted parameters shown in Table 3.5 did not differ much from the previously obtained values (see Table 3.4).

The fact that the adjusted value of transcapillary rate of escape of globulins (TER_{Glob}) is much closer to the adjusted value for albumin (TER_{Alb}), compared to the values set for the original model, can be probably explained by the fact that both in the model and in the data globulins represent a wide range of non-albumin proteins including those with molecular weight and dimensions lower than that of albumin (e.g. α1-globulin), whereas the original value of TER_{Glob} was assumed for an average globulin based on the assumed average molecular weight of globulins.

Note that the aim of the fitting procedure presented in this study was not to estimate specific parameters, but to show that the proposed model is able to match clinical data with reasonable accuracy, following adjustments in selected model parameters, thus confirming the flexibility of the model. Therefore, no specific

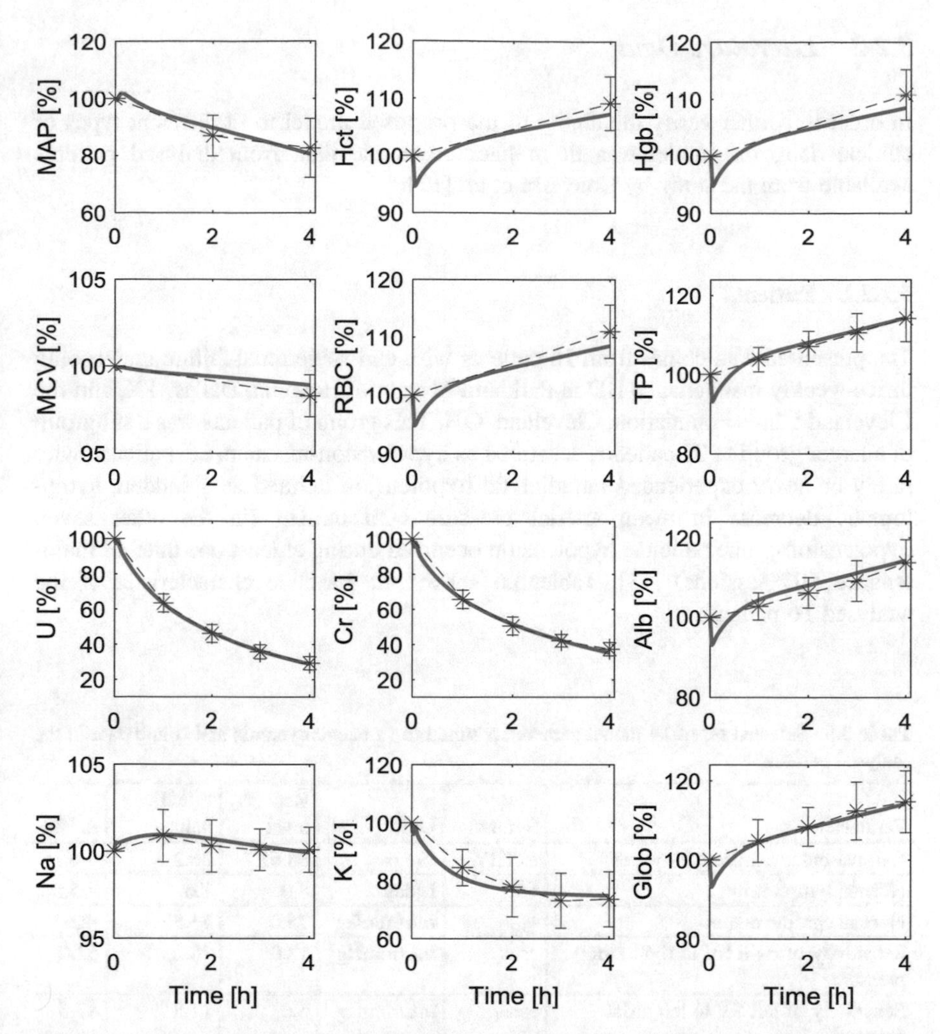

Fig. 3.5 Percentage changes in the main haemodynamic parameters and plasma solute concentrations during a 4h HD procedure – model simulations (solid lines) and in vivo data from patients from Lublin Medical University, Poland (asterisks connected with dashed lines). MAP, mean blood pressure in large arteries; Hct, haematocrit; RBC, red blood cell count; Hgb, haemoglobin; MCV, mean corpuscular volume of erythrocytes. All blood data corresponds to blood in the arterial dialysis tubing

conclusions are drawn regarding individual parameters, for which more data and a more deep and accurate fitting procedure should be employed. Ideally, the data from more patients, from several dialysis sessions in the same patient and with more solute and haemodynamic data recorded, would be needed for a meaningful analysis of parameter values.

3.2.2 Literature Data

In order to further verify the ability of the proposed model to fit different types of clinical data, the model was fit to haemodynamic data from dialysed patients available from the study by Converse et al. [13].

3.2.2.1 Patients

The presented data comes from 16 patients with end-stage renal failure undergoing thrice-weekly maintenance HD in Parkland Memorial Hospital, Dallas, TX, and the Cleveland Clinic Foundation, Cleveland, OH. This group of patients was a subgroup of a larger group of 23 patients, described as hypotension-resistant, i.e. patients, who rarely or never experienced intradialytic hypotension defined as a sudden, symptomatic decrease in mean arterial pressure >30 mmHg (in the other seven hypotension-prone patients, hypotension occurred during at least one third of maintenance HD sessions) [13]. Table 3.6 shows the baseline characteristic of the analysed 16 patients.

Table 3.5 Selected set of 14 model parameters tuned to fit haemodynamic and solute data in the analysed patients

Parameter	Symbol	Unit	Ref. level	Fitted value	Δ [%]
Extravascular extracellular water	exTBW	%	38.6‡	36.2	−6.1
Normal lymph flow	QL_n	L/day	8.0	7.6	−5.5
Normal cardiac output	q_{kg}	mL/min/kg	75.0	73.3	−2.2
Sensitivity of right SV to right atrial pressure	$sens_r$	mL/mmHg	12.0	18.2	52.0
Sensitivity of left SV to left atrial pressure	$sens_l$	mL/mmHg	6.0	11.3	87.6
Transcapillary escape rate of albumin	TER_{Alb}	%/h	5.0	5.6	11.2
Transcapillary escape rate of globulins	TER_{Glob}	%/h	3.0	4.3	44.7
Albumin capillary reflection coefficient	$\sigma_{alb,cap}$	–	0.95	0.99	4.0
Transcellular urea mass transfer	$K_{U,cell}$	L/min	0.80	0.20	−75.0
Transcellular creatinine mass transfer	$K_{Cr,cell}$	L/min	0.60	0.18	−70.3
Arterioles resistance amplitude	ΔR	mmHg*s/mL	1.0	0.56	−44.0
Venous unstressed volume amplitude	ΔV	mL	600	942	57.0
Arterial vagal gain of T regulation	$G_{a,Tv}$	s/mmHg	0.02	0.02	−25.0
Cardiopulmonary gain of $V_{u,sv}$ regulation	$G_{c,V}$	mL/mmHg	208	0.4	−99.8

‡ Estimated from Ref. [11].

3.2.2.2 Data

The following haemodynamic parameters were measured before and during dialysis: arterial pressure was measured continuously using the Finapres® device (mean arterial pressure was calculated as one third of the pulse pressure plus diastolic pressure), heart rate was measured continuously using electrocardiography, and the calf blood flow was recorded once every 15 s using venous-occlusion plethysmography [13]. Calf vascular resistance was then calculated as the mean arterial pressure divided by calf blood flow. The baseline measurements were recorded over a 15-minute period before dialysis when the patient remained stable in the supine position. Four data points were reported in the cited article corresponding to 0%, 33%, 67% and 100% of dialysis duration [13]. For the purpose of the fitting procedure, it was assumed that the changes in calf blood flow and calf vascular resistance correspond to the changes in cardiac output and total systemic vascular resistance (similarly as assumed by Ursino in his study [14]), and hence the simulated (normalised) cardiac output and systemic resistance were fitted to the normalised data on calf blood flow and resistance.

3.2.2.3 Fitting Procedure

The fitting procedure was similar to that described in the previous chapter using the modified Matlab function *fminsearchbnd* with all data normalised by the baseline values. The quality of fit was again measured as the sum of the squared deviations between the simulated variables and clinical data for all available 12 data points (3 for each variable). Before simulating the HD, the modelled system was brought to the initial steady state corresponding to the baseline data in the analysed group of patients (see Table 3.6).

For the HD simulation, the average dialysis settings in the group of analysed patients were used in the model, as listed in Table 3.7 (for the blood flow rate the middle value of the provided range was used). Since the data on plasma concentrations of solutes other than sodium and potassium was not available, their initial concentration in both plasma and dialysate fluid were left as in the original model for the reference healthy subject. Based on data on pre- vs post-dialysis weight

Table 3.6 Characteristics of 16 patients from the Converse study (mean ± SE) [13]

Age	43 ± 3	years
Weight before dialysis	80 ± 6	kg
Plasma sodium	137 ± 1	mmol/L
Plasma potassium	5.0 ± 0.2	mmol/L
Plasma urea	25 ± 3	mmol/L
Plasma creatinine	1.4 ± 0.1	mmol/L
Haematocrit	27 ± 1	%
Mean arterial pressure	105 ± 5	mm Hg
Heart rate	74 ± 4	bpm

Table 3.7 Dialysis settings in the group of 16 patients from the Converse study [13]

Ultrafiltration	2.8	L
Dialysate flow	500	mL/min
Blood flow rate	300–400	mL/min
Dialysate sodium	142	mmol/L
Dialysate potassium	2	mmol/L

difference and the dialyzer ultrafiltration, it was assumed that the priming saline was discarded and not infused to the patients.

Twenty-seven parameters were chosen to be fitted (see Table 3.8). These were most of the parameters identified by the sensitivity analysis from Sect. 3.1, the gains and amplitudes of all baroreflex mechanisms and a few other parameters related to the cardiovascular system and the lymphatic system. In general no parameters related to solutes were adjusted in this fitting procedure, given the lack of solute data, with the exception of parameters related to transcapillary protein transport, which affect significantly the vascular refilling mechanism and the blood volume during dialysis and hence affect the blood pressure and baroreflex mechanisms. All other parameters were left at the values set in the original model for the reference healthy subject.

3.2.2.4 Fitting Results

As shown in Fig. 3.6, the model can fit relatively well the clinical data. The obtained global level of fit was 30 percentage point squared, which gives an average error of circa 1.6% for each data point. The analysed patients showed on average a very mild pressure decrease, a steady increase in the vascular resistance, a gradual decrease of cardiac output and very moderate changes in the heart rate. These patterns were obtained in the model by some substantial changes in the parameters of the baroreflex mechanisms and relatively moderate changes in the remaining parameters (within the expected range of physiological variability), as listed in Table 3.8. As far as the baroreflex parameters are concerned, the most important changes in the parameters values included:

- An increase in the capacity (amplitude) of the mechanism controlling vascular resistance
- A decrease in the ratio of cardiopulmonary and arterial gains of the mechanism controlling vascular resistance (a change towards equalising the relative importance of both group of baroreceptors)
- A decrease of the importance of the mechanism controlling heart rate (decrease of gains from both groups of baroreceptors)
- An increase of both cardiopulmonary and arterial gains of the mechanism controlling venous unstressed volume
- A decrease of the arterial gain of the mechanism controlling heart contractility

Table 3.8 Model parameters tuned to fit haemodynamic data from the Converse study

Parameter	Symbol	Unit	Ref. level	Fitted value	Δ [%]
Extravascular extracellular water	exTBW	%	35.7	36.0	0.8
Normal cardiac output	q_{kg}	mL/min/kg	75.0	76.6	2.2
Sensitivity of right SV to right atrial pressure	$sens_r$	mL/mmHg	12.0	14.4	19.7
Sensitivity of left SV to left atrial pressure	$sens_l$	mL/mmHg	6.0	5.9	−2.2
Normal lymph flow	QL_n	L/day	8.0	5.5	−30.7
Capillary hydraulic conductivity	Lp	mL/min/mmHg	4.5	4.0	−11.1
Transcapillary escape rate of albumin	TER_{Alb}	%/h	5.0	6.4	28.8
Transcapillary escape rate of globulins	TER_{Glob}	%/h	3.0	3.4	13.2
Albumin capillary reflection coefficient	$\sigma_{alb,cap}$	–	0.95	0.98	3.2
Lymph flow sensitivity to P_{is}	LS	mL/mmHg/h	43.0	45.3	5.4
Interstitial volume excluded to proteins	$V_{is,ex,\%}$	%	50.0	53.2	6.4
Interstitial compliance	C_{is}	% of $V_{is,n}$/mmHg	12.0	14.4	20.2
Plasma water fraction	F_{pl}	–	0.94	0.95	0.6
Ratio of whole-body HCT to central HCT	F_{cells}	–	0.90	0.92	1.7
Dialyzer Gibbs-Donnan coefficient for sodium	α_{DNa}	–	0.94	0.95	0.5
Heart period amplitude	ΔT	s	0.70	0.77	10.3
Arterioles resistance amplitude	ΔR	mmHg*s/mL	1.0	1.6	61.9
Venous unstressed volume amplitude	ΔV	mL	600	636	6.0
Heart contractility amplitude	ΔE	–	0.50	0.58	16.8
Arterial vagal gain of T regulation	$G_{a,Tv}$	s/mmHg	0.02	0.005	−73.0
Arterial sympathetic gain of T regulation	$G_{a,Ts}$	s/mmHg	0.01	0.004	−64.0
Cardiopulmonary gain of R regulation	$G_{c,R}$	s/mL	0.63	0.11	−82.0
Arterial gain of R regulation	$G_{a,R}$	s/mL	0.03	0.08	159.7
Cardiopulmonary gain of $V_{u,sv}$ regulation	$G_{c,V}$	mL/mmHg	208	304	46.4
Arterial gain of $V_{u,sv}$ regulation	$G_{a,V}$	mL/mmHg	14	23	63.3
Cardiopulmonary gain of E regulation	$G_{c,E}$	1/mmHg	0.15	0.15	47.7
Arterial gain of E regulation	$G_{a,E}$	1/mmHg	0.01	0.00	−94.0

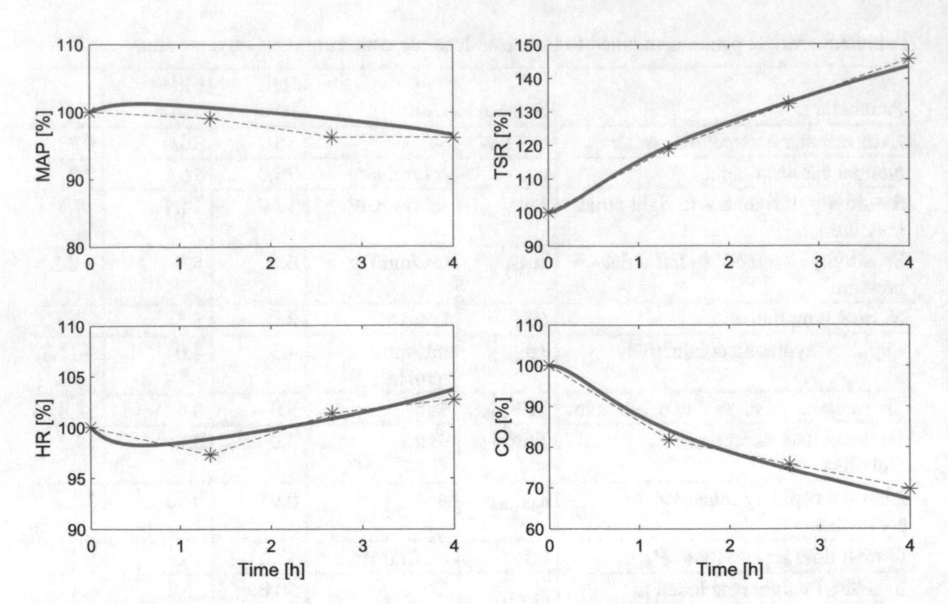

Fig. 3.6 Percentage changes in the main haemodynamic quantities during a 4 hours haemodialysis – model simulations (solid lines) and in vivo data averaged for 16 patients studied by Converse et al. [13] (asterisks connected with dashed lines). MAP, mean pressure in large arteries; HR, heart rate; TSR, total systemic resistance (calf resistance for in vivo data); CO, cardiac output (calf blood flow for in vivo data)

3.3 Model Limitations

Due to some of the assumptions and simplifications made (necessary to keep the model complexity at a reasonable level), the proposed model is subject to certain limitations.

First of all, because of the absence of the time-varying elastance model of the heart, the cardiac output is not pulsatile, and hence, the model can simulate only mean blood pressure changes. Given that the baroreceptors normally respond also to changes in the pulse pressure [15], the lack of cardiac pulsatility distorts slightly the simulation results at times when considerable haemodynamic changes occur within a short period of time, e.g. during and immediately after filling of the extracorporeal circuit with the patient's blood. This phenomenon is, however, rather negligible for the simulation of haemodialysis, given its relatively long time scale (in the range of hours). As shown in the previously developed model of the Valsalva manoeuvre [16, 17], the proposed non-pulsatile model was sufficient even for simulating MAP variations over time intervals of seconds corresponding to a few heart beats.

Moreover, the baroreceptors in the model respond only to changes in MAP while not being sensitive to the rate of these changes. The lack of a rate-dependent component in the baroreflex mechanism can have an impact on the short-term (~1 s) transient haemodynamic response to the given perturbation; however, it should not affect the system's state in the longer term (the influence of the pressure time derivatives could be greater in a model involving pulsating blood pressure [18]).

Also, the baroregulation of heart rate in the model combines the activity of sympathetic and parasympathetic systems; however, it does not account for their possible negative mutual interactions (see the discussion on the negative sympathetic-parasympathetic interaction in [18]).

The model does not account for the possible resetting of baroreceptors, i.e. the wane of baroreflex afferent and efferent activity despite sustained changes in blood pressure. Such resetting may occur both in the short-term (acute baroreceptor resetting over the course of seconds to minutes) and in the long-term (chronic baroreceptor resetting over the course of days to weeks) [19]. There is an ongoing debate on the scope and mechanisms of such resetting and whether the baroreflex is involved in the long-term control of arterial blood pressure [19]. On one hand, there is evidence that chronic electrical stimulation or carotid baroreceptor afferent nerve fibres in dogs leads to a decrease in arterial pressure and a decrease in sympathetic activity sustained for over 7 days [20]. Analogically, chronic unloading of carotid baroreceptors in dogs causes an increase in arterial pressure sustained for over 7 days [21]. Similarly, a sustained increase in arterial blood pressure caused by infusion of Angiotensin II for a week results in sustained reduction of renal sympathetic nerve activity (RSNA) [22, 23]. These experiments suggest that baroreceptors do not reset and can contribute to the chronic control of arterial BP and RSNA. On the other hand, there is evidence that arterial baroreceptors (particularly myelinated A-fibres) reset to sustained changes in arterial pressure and that the arterial pressure remains unchanged following baroreceptor deafferentation, suggesting that the baroreflex cannot be involved in the long-term control of BP [19]. A possible explanation suggests that the myelinated A-fibres are subject to resetting, while unmyelinated C-fibres do not reset or reset to a lesser degree, and that it is the activity of the latter, which maintains changes in sympathetic activity and arterial pressure [19]. Moreover, chronic resetting of baroreceptors is usually incomplete [24–26], although a complete resetting was reported for the baroreflex control of heart rate in hypertension and pregnancy [19] and in control of RSNA in renal hypertension [27, 28], suggesting that the sympathetic activity to individual organs is controlled differentially [19]. It is also possible that the slope (gain) of the baroreflex afferent activity changes with sustained changes in blood pressure [29]. The possible acute resetting of baroreceptors may obviously impact the cardiovascular response to HD, whereas chronic baroreceptor resetting may have an impact on the long-term arterial BP due to chronic fluid overload of dialysis patients. Both phenomena are complex and incompletely understood and therefore are beyond the scope of this study.

The model does not distinguish between neural and humoral regulation (e.g. renin-angiotensin or vasopressin), and hence, all regulation in the model is due solely to the baroreflex. Similarly, the model ignores the regulatory mechanisms induced by the activity of osmoreceptors, chemoreceptors or pulmonary stretch receptors [30, 31] and does not take into account the cardiopulmonary interactions (such as the impact of lung inflation on the blood flow through pulmonary vessels) or the gas exchange at the level of pulmonary capillaries.

As is clear from the structure of the cardiovascular model, in which all systemic circulation is lumped together, the model does not account for the differences in terms of vascular properties, haematocrit level or solute kinetics between different (regional) blood circulations [32, 33], and it does not take into account the distinct behaviour of different parts of the venous system or the autoregulation of individual organs (especially the brain, heart and kidneys), all of which can affect the arterial blood pressure during HD. For instance, changes in the resistance of systemic arterioles can be induced by both baroreflex regulation and local vascular changes, with both mechanisms acting either in synergism or in antagonism [34].

The transcapillary water and solute transport modelled in the systemic capillaries compartment describes the whole-body transcapillary transport processes, including those taking place at the level of pulmonary capillaries. Similarly, the description of lymph absorption from the interstitial space reflects the whole-body lymph flow, including the lymph absorbed from the pulmonary interstitial fluid. This is, of course, a simplification since the properties of the pulmonary capillary walls and the pulmonary interstitial fluid are different compared to those of other tissues [35]. In particular, the hydraulic pressure in the capillary walls is lower than in the systemic capillaries, the pulmonary capillaries are more permeable to proteins, and the hydrostatic pressure of the pulmonary interstitial fluid is more negative than that of the interstitial fluid in the peripheral tissues [35]. For a more accurate description of the transcapillary filtration/absorption of fluid, the transcapillary protein leakage and protein refilling through the flow of lymph, these processes should be ideally modelled separately for the systemic and pulmonary circulations.

Moreover, it was assumed in the model that the properties of capillary and cellular membranes (i.e. solutes permeability, solutes reflection coefficients and hydraulic conductivity) are the same for both sides of each membrane. It was also assumed that these properties remain constant and do not depend on the solute concentrations on either side of the membrane or on the transmembrane potential (not considered in the model). Similarly, it was assumed that the parameters describing uneven distribution of ions across the cellular membrane (β) are constant and independent of ion concentration. During haemodialysis, however, the properties of all membranes can potentially change, thus influencing the analysed transport processes.

The model does not account for the acid-base status of the patient or chemical reactions describing the buffer systems; it neglects protein binding of solutes, and it ignores the temperature effects of dialysis.

Finally, the model does not account for the possible occurrence of the sympatho-inhibitory or cardio-inhibitory reflex (the Bezold-Jarisch reflex [36, 37]), i.e. a sudden withdrawal of the sympathetic drive and an increase of the parasympathetic activity leading to a sudden decrease in arterial blood pressure and bradycardia in response to severe hypovolaemia or myocardial ischaemia. This, seemingly paradoxical, reflex is thought to play a cardio-protective role, when cardiac filling is severely reduced [36, 37].

The above model limitations can be addressed in future, more complex versions of the model, approaching the philosophy of integrative modelling of human physiology [38, 39].

References

1. Kyriazis, J., Kalogeropoulou, K., Bilirakis, L., Smirnioudis, N., Pikounis, V., Stamatiadis, D., Liolia, E.: Dialysate magnesium level and blood pressure. Kidney Int. **66**(3), 1221–1231 (2004)
2. Rothe, C.: Reflex control of veins and vascular capacitance. Physiol Rev. **63**(4), 1281–1342 (1983)
3. Debowska, M., Poleszczuk, J., Wojcik-Zaluska, A., Ksiazek, A., Zaluska, W.: Phosphate kinetics during weekly cycle of hemodialysis sessions: application of mathematical modeling. Artif Organs. **39**(12), 1005–1014 (2015)
4. Poleszczuk, J., Debowska, M., Wojcik-Zaluska, A., Ksiazek, A., Zaluska, W.: Phosphate kinetics in hemodialysis: application of delayed pseudo one-compartment model. Blood Purif. **42**(3), 177–185 (2016)
5. Debowska, M., Wojcik-Zaluska, A., Ksiazek, A., Zaluska, W., Waniewski, J.: Phosphate, urea and creatinine clearances: haemodialysis adequacy assessed by weekly monitoring. Nephrol Dial Transplant. **30**, 129–136 (2015)
6. Waniewski, J., Debowska, M., Wojcik-Zaluska, A., Ksiazek, A., Zaluska, W.: Quantification of dialytic removal and extracellular calcium mass balance during a weekly cycle of hemodialysis. PLoS One. **11**(4), e0153285 (2016)
7. Pietribiasi, M., Waniewski, J., Załuska, A., Załuska, W., Lindholm, B.: Modelling transcapillary transport of fluid and proteins in hemodialysis patients. PLoS One. **11**(8), e0159748 (2016)
8. Pietribiasi, M., Waniewski, J., Wojcik-Zaluska, A., Zaluska, W., Lindholm, B.: Model of fluid and solute shifts during hemodialysis with active transport of sodium and potassium. PLoS ONE. **13**(12), e0209553 (2018)
9. Siemens Healthcare Diagnostics Inc.: ADVIA® 2120/2120i Hematology Systems Operator's Guide, Rev. C, 2010-04. Siemens Healthcare Diagnostics Inc., Tarrytown, NY (2010)
10. Hume, R., Weyers, E.: Relationship between total body water and surface area in normal and obese subjects. J Clin Pathol. **24**(3), 234–236 (1971)
11. Bhave, G., Neilson, E.: Body fluid dynamics: back to the future. J Am Soc Nephrol. **22**, 2166–2181 (2011)
12. Magosso, E., Biavati, V., Ursino, M.: Role of the baroreflex in cardiovascular instability: a modeling study. Cardiovasc Eng. **1**(2), 101–115 (2001)
13. Converse Jr., R., Jacobsen, T., Jost, C., Toto, R., Grayburn, P., Obregon, T., Fouad-Tarazi, F., Victor, R.: Paradoxical withdrawal of reflex vasoconstriction as a cause of hemodialysis-induced hypotension. J Clin Invest. **90**(5), 1657–1665 (1992)
14. Ursino, M., Innocenti, M.: Modeling arterial hypotension during hemodialysis. Art Org. **21**(8), 873–890 (1997)
15. Looga, R.: The Valsalva manoeuvre—cardiovascular effects and performance technique: a critical review. Respir Physiol Neurobiol. **147**, 39–49 (2005)
16. Pstras, L., Thomaseth, K., Waniewski, J., Balzani, I., Bellavere, F.: Mathematical modelling of cardiovascular response to the Valsalva manoeuvre. Math Med Biol. **34**(2), 261–292 (2017)
17. Pstras, L., Thomaseth, K., Waniewski, J., Balzani, I., Bellavere, F.: Modeling pathological hemodynamic responses to the Valsalva maneuver. J Biomech Eng. **139**(6), 061001-1-9 (2017)
18. Ursino, M.: Modelling the interaction among several mechanisms in the short-term arterial pressure control. In: Mathematical Modelling in Medicine, pp. 139–161. IOS Press, Amsterdam (2000)
19. Brooks, V., Sved, A.: Pressure to change? Re-evaluating the role of baroreceptors in the long-term control of arterial pressure. Am J Physiol Regul Integr Comp Physiol. **288**(4), R815–R818 (2005)
20. Lohmeier, T., Irwin, E., Rossing, M., Serdar, D., Kieval, R.: Prolonged activation of the baroreflex produces sustained hypotension. Hypertension. **43**(2), 306–311 (2004)
21. Thrasher, T.: Unloading arterial baroreceptors causes neurogenic hypertension. Am J Physiol Regul Integr Comp Physiol. **282**(4), R1044–R1053 (2002)

22. Barrett, C., Ramchandra, R., Guild, S., Lala, A., Budgett, D., Malpas, S.: What sets the long-term level of renal sympathetic nerve activity: a role for angiotensin II and baroreflexes? Circ Res. **92**(12), 1330–1336 (2003)
23. Lohmeier, T., Lohmeier, J., Haque, A., Hildebrandt, D.: Baroreflexes prevent neurally induced sodium retention in angiotensin hypertension. Am J Physiol Regul Integr Comp Physiol. **279**(4), R1437–R1448 (2000)
24. Munch, P., Andresen, M., Brown, A.: Rapid resetting of aortic baroreceptors in vitro. Am J Physiol. **244**(5), H672–H680 (1983)
25. Lohmeier, T., Lohmeier, J., Warren, S., May, P., Cunningham, J.: Sustained activation of the central baroreceptor pathway in angiotensin hypertension. Hypertension. **39**(2 Pt 2), 550–556 (2002)
26. Thrasher, T.: Effects of chronic baroreceptor unloading on blood pressure in the dog. Am J Physiol Regul Integr Comp Physiol. **288**(4), R863–R871 (2005)
27. Bell, L., Wilson, D., Quandt, L., Kampine, J.: Renal sympathetic and heart rate baroreflex function in conscious and isoflurane anaesthetized normotensive and chronically hypertensive rabbits. Clin Exp Pharmacol Physiol. **22**(10), 701–710 (1995)
28. Head, G., Burke, S.: Renal and cardiac sympathetic baroreflexes in hypertensive rabbits. Clin Exp Pharmacol Physiol. **28**(12), 972–975 (2001)
29. Andressen, M., Yang, M.: Arterial baroreceptor resetting: contributions of chronic and acute processes. Clin Exp Pharmacol Physiol. **Suppl 15**, 19–30 (1989)
30. Looga, R.: The bradycardic response to the Valsalva manoeuvre in normal man. Respiration Physiology. **124**, 205–215 (2001)
31. Junqueira Jr., L.: Teaching cardiac autonomic function dynamics employing the Valsalva. Adv Physiol Educ. **32**, 100–106 (2008)
32. Mchedlishvili, G., Varazashvili, M.: Hematocrit in cerebral capillaries and veins under control and ischemic conditions. J Cereb Blood Flow Metab. **7**(6), 739–744 (1987)
33. Schneditz, D., Platzer, D., Daugirdas, J.: A diffusion-adjusted regional blood flow model to predict solute kinetics during haemodialysis. Nephrol Dial Transplant. **24**, 2218–2224 (2009)
34. Ursino, M., Antonucci, M., Belardinelli, E.: Role of active changes in venous capacity by the carotid baroreflex: analysis with a mathematical model. Am J Physiol Heart Circ Physiol. **267**(6), H2531–H2546 (1994)
35. Guyton, A., Hall, J.: Textbook of medical physiology, 11th edn. Elsevier Saunders, Philadelphia (2006)
36. Campagna, J., Carter, C.: Clinical relevance of the Bezold-Jarisch reflex. Anesthesiology. **98**(5), 1250–1260 (2003)
37. Kashihara, K., Kawada, T., Yanagiya, Y., Uemura, K., Inagaki, M., Takaki, H., Sugimachi, M., Sunagawa, K.: Bezold–Jarisch reflex attenuates dynamic gain of baroreflex neural arc. Am J Physiol Heart Circ Physiol. **285**(2), H833–H840 (2003)
38. Coleman, T., Randall, J.: HUMAN. A comprehensive physiological model. Physiologist. **26**(1), 15–21 (1983)
39. Hester, R., Brown, A., Husband, L., Iliescu, R., Pruett, D., Summers, R., Coleman, T.: HumMod: a modeling environment for the simulation of integrative human physiology. Front Physiol. **2**, 12 (2011)

Chapter 4
Computational Simulations of Patient's Response to Fluid and Solute Removal by Haemodialysis

Abstract In this chapter model simulations for the reference virtual patient are used to discuss the human's body response to water and solute removal during a standard haemodialysis treatment and to describe the nature and magnitude of physiological processes taking place within the cardiovascular system and extravascular fluid compartments during a dialysis session. The impact of dialysis settings and baroreflex or cardiac parameters on patient's blood pressure response to haemodialysis is studied indicating some of the possible mechanisms of intradialytic hypo- or hypertension.

Keywords Model simulations · Haemodynamics · Mean arterial blood pressure · Intradialytic hypotension · Intradialytic hypertension · Priming saline · Vascular refilling · Fluid absorption · Lymph flow · Solute kinetics · Haematocrit · Baroreflex parameters · Stroke volume · Atrial pressure · Filling pressure · Dialysis fluid composition · Ultrafiltration

4.1 Haemodynamics

4.1.1 Case 1 (Priming Saline Infused into the Patient)

As shown in Fig. 4.1, when the priming saline is not discarded, the reference subject should tolerate the removal of 3.2 L of fluid during a 4 hours HD session with almost no arterial hypotension. Despite the blood volume reduction by over 12% (in particular plasma volume reduction by almost 18%), the mean arterial blood pressure decreases at the end of dialysis only by circa 3%. This is achieved by the operation of the baroreflex mechanisms – a strong increase in systemic resistance (+44%), a strong reduction of venous unstressed volume (−45%) and a moderate increase in heart contractility (+18%).

As far as the heart rate is concerned, at the end of dialysis session, the heart rate is only slightly higher than before dialysis (+5%), which is an expected result for the blood volume reduction of over 10% [1]. During dialysis, however, the heart rate

L. Pstras, J. Waniewski, *Mathematical Modelling of Haemodialysis*,
https://doi.org/10.1007/978-3-030-21410-4_4

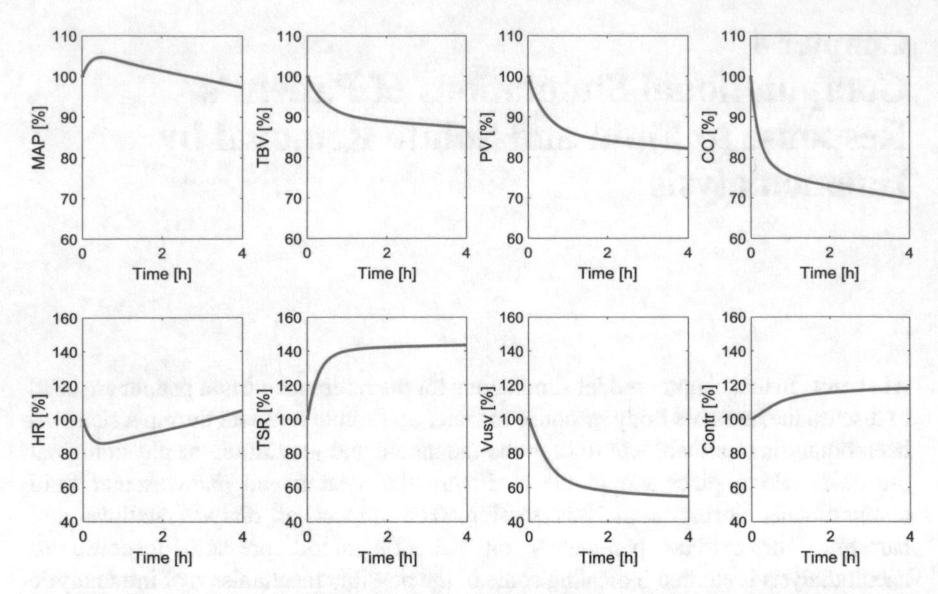

Fig. 4.1 Simulated changes in the main haemodynamic variables during haemodialysis (Case 1: the priming saline infused to the patient). MAP, mean blood pressure in large arteries; TBV, total blood volume; PV, plasma volume; CO, cardiac output; HR, heart rate; TSR, total systemic resistance; Vusv, systemic veins unstressed volume; Contr, heart contractility

follows the characteristic biphasic pattern seen in fluid-overloaded patients on HD, i.e. the initial drop in HR at the first phase of dialysis (when the blood volume drops relatively rapidly) and a subsequent increase in HR during progressive (slower) reduction in the blood volume [1].

Despite the continuous increase in heart contractility and the later increase in heart rate, the cardiac output decreases substantially (−30%), which is due to the largely reduced cardiac preload (the right atrial pressure), being only partly compensated by the slightly reduced afterload (the pressure in the large arteries).

As shown in Fig. 4.1, in the first part of the simulated dialysis session, the arterial pressure increases slightly over the initial baseline level. This paradoxical pressure increase can be explained by the prevalence of the activity of cardiopulmonary baroreceptors over the arterial baroreceptors (see the gains of the baroreflex mechanisms in Table 2.3 assigned for the normal subject) [2, 3]. Similar patterns of arterial BP during HD were reported in the literature [2], although it is not necessarily a norm for dialysis patients, who, depending on the activity of baroreflex and other regulatory mechanisms, may present various blood pressure responses, such as a decrease of pressure from the beginning of dialysis or a continuous increase of pressure throughout the dialysis session. The above phenomenon is governed mainly by the baroreflex mechanism controlling the vascular resistance discussed in more detail in Sect. 4.6 (please see Fig. 4.13 showing the impact of individual baroreflex mechanisms and Fig. 4.16 showing how decreasing the cardiopulmonary gain of this mechanism changes the profile of blood pressure changes during HD).

Please note that Fig. 4.1 shows only the simulation for the duration of the dialysis itself, without accounting for the pre-dialysis procedure of filling the extracorporeal circuit with the patient's blood and the subsequent short 'idle' time before starting the dialysis. The latter is shown in Fig. 4.2, where one can see the initial increase in the total blood volume and total plasma volume following the infusion of the priming saline (for all simulations presented in this study, the total blood volume refers to the total blood circulating in the whole system, thus including the blood circulating outside the body within the dialyzer circuit). When the whole procedure is considered, the total blood volume is reduced by circa 9%, while the total plasma volume is reduced by circa 13%. The fact that during the dialysis alone (see Fig. 4.1), the total blood volume is reduced by over 12%, despite the higher initial level of reference, is due to the fact that infusing the saline into the circulation has a diluting effect on the blood, and hence part of the saline is quickly filtered out of the vascular space to the interstitial fluid, thus restoring the osmotic (oncotic) balance across the capillary walls.

As can be seen in Fig. 4.2, choosing the exact moment of beginning the measurements of blood volume changes can have an impact on the observed values. This becomes even more important in smaller patients with low blood volume, in whom the volume of the infused saline will be proportionally larger (for this reason smaller dialyzers are used in paediatric patients).

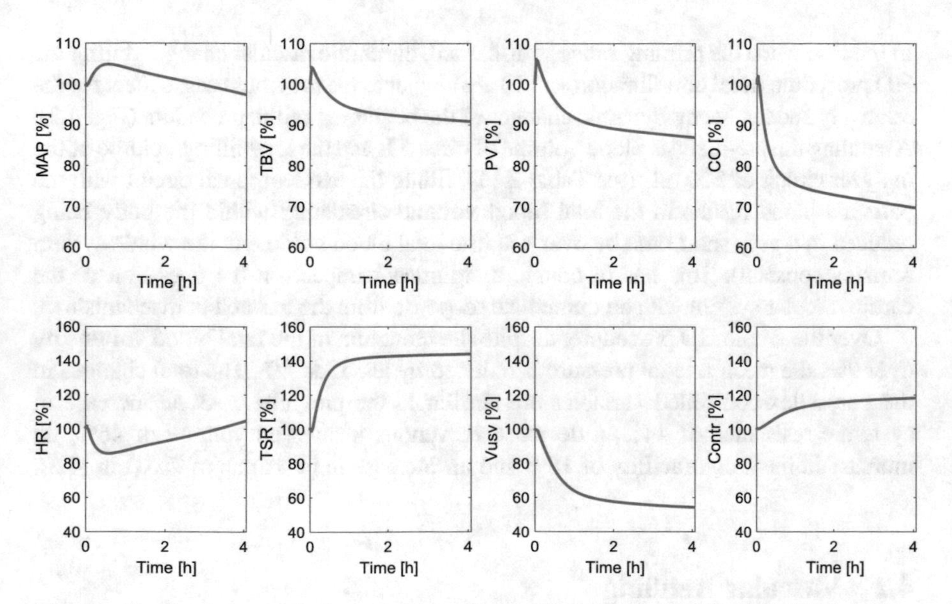

Fig. 4.2 Simulated changes in the main haemodynamic variables during the whole HD procedure including filling of the extracorporeal circuit with blood (Case 1: the priming saline infused to the patient)

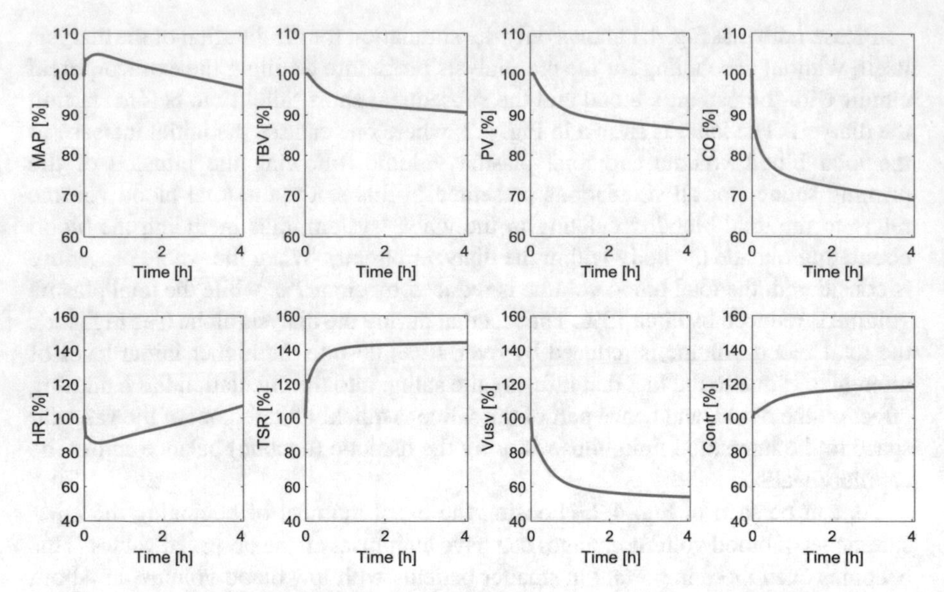

Fig. 4.3 Simulated changes in the main haemodynamic variables during the whole HD procedure including filling of the extracorporeal circuit with blood (Case 2: the priming saline discarded)

4.1.2 Case 2 (Priming Saline Discarded)

In the case when the priming saline is discarded, the cardiovascular changes during the HD procedure (total ultrafiltration = 3 L) are similar to the previous case, except for the relatively sudden haemodynamic changes at the beginning of the procedure (Fig. 4.3). Assuming the pre-dialysis blood volume of circa 5 L and the total filling volume of the dialyzer circuit of 220 mL (see Table 2.11), filling the extracorporeal circuit with the patient's blood results in the total blood volume circulating within the body being reduced in a very short time by over 4% (the total blood volume in the whole system remains constant). This has of course a significant impact on the operation of the cardiovascular system with an immediate response from the baroreflex mechanisms.

Over the whole HD procedure, despite the reduction of the total blood volume by over 9%, the mean arterial pressure is reduced by less than 4%. The total changes in the baroreflex-controlled variables are similar to the previous case: an increase in systemic resistance of 44%, a decrease in venous unstressed volume of 46%, an increase in heart contractility of 18% and an increase in heart rate of 7% (Fig. 4.3).

4.2 Vascular Refilling

Haemodynamic changes during HD reflect not only the operation of the baroreflex mechanisms but also the so-called vascular refilling – the process of refilling the vascular system with the fluid from the tissues to maintain the blood volume and enable a relatively stable operation of the cardiovascular system. This mechanism takes two pathways: fluid absorption from the interstitium across the capillary walls and lymph absorption from the interstitium via the lymphatic system (the lymph is eventually drained to the systemic veins) [4, 5].

Under normal physiological conditions, in the vast majority of tissues, there is a net filtration of fluid (water and solutes) out of the capillaries to the interstitium [6, 7] dictated by a small imbalance between the hydraulic pressure difference and osmotic (mainly oncotic) pressure difference across the capillary wall (a similar case applies for the fluid overload state; see the slightly positive total net pressure at time 0 in Fig. 4.4 representing the reference patient before dialysis procedure). Under steady-state conditions, the magnitude of capillary filtration equals the total lymph flow in the lymphatic system, and hence the total blood volume remains constant (vascular refilling equal to 0, see the last panel in Fig. 4.5 at time 0). When the capillary blood pressure starts to decrease during haemodialysis (as the result of blood volume reduction), the filtration out of the capillaries is attenuated due to the reduced pressure difference between the capillaries and the surrounding interstitium. At the same time, the oncotic pressure difference between capillary plasma and interstitium starts to increase (following the initial drop due to dilution of blood by the saline infusion). As the capillary blood pressure continues to decline, the capillary filtration keeps decreasing, until the fluid filtration ceases and reverses to 'negative' values indicating fluid absorption from the tissues across the capillary walls (see Fig. 4.5). From this point on, the fluid is absorbed from the tissues by both the lymphatic system and the capillaries.

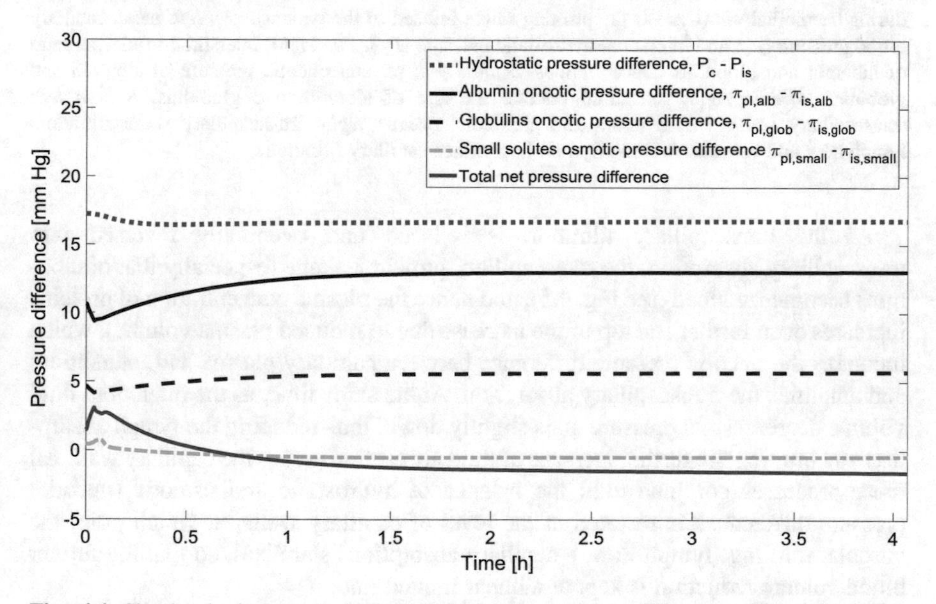

Fig. 4.4 Simulated changes in hydrostatic/hydraulic pressure difference, albumin/globulins oncotic pressure difference and small solute osmotic pressure difference between the capillary plasma and interstitial fluid during the HD procedure (Case 1: the priming saline infused to the patient)

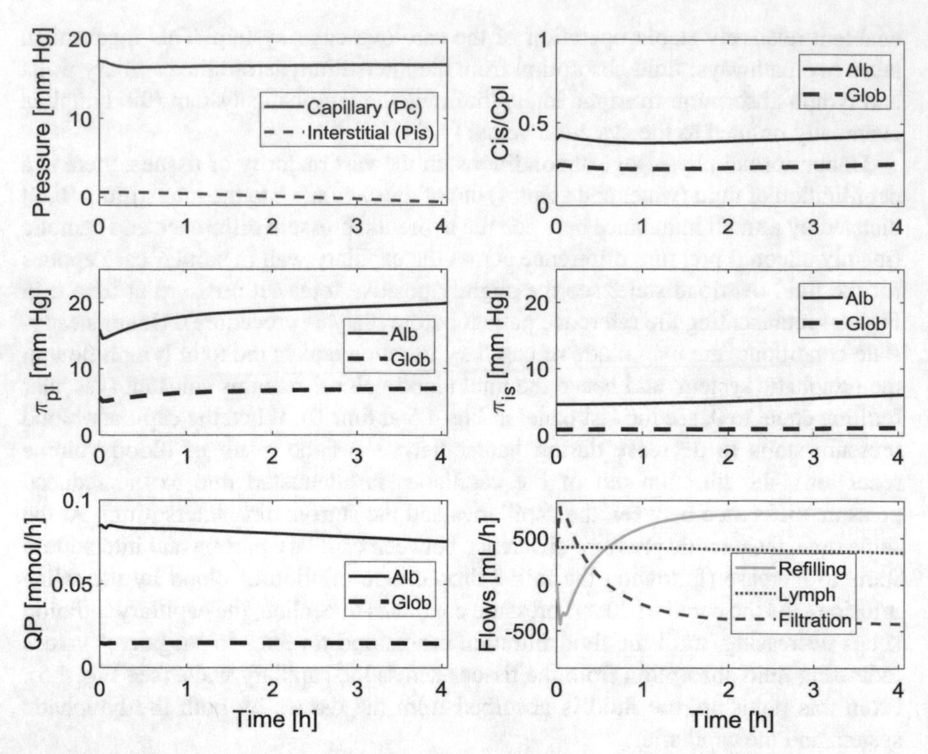

Fig. 4.5 Simulated changes in the variables describing vascular refilling and protein transport during haemodialysis (Case 1: the priming saline infused to the patient). *Top left*: mean capillary blood pressure (P_c) and interstitial hydrostatic pressure (P_{is}). *Top right*: interstitial-to-plasma ratio of albumin and globulins concentrations. *Middle left*: plasma oncotic pressure of albumin and globulins. *Middle right*: interstitial oncotic pressure of albumin and globulins. *Bottom left*: transcapillary leakage of albumin and globulins. *Bottom right*: transcapillary water filtration, lymph flow and vascular refilling (lymph flow minus capillary filtration)

As the transcapillary filtration is reduced and eventually reversed into transcapillary absorption, the transcapillary protein leakage (especially that of albumin) becomes reduced (see Fig. 4.5), and hence the plasma concentration of proteins increases even further (on top of the increase due to reduced plasma volume), which increases the oncotic pressure difference between capillary plasma and interstitium and amplifies the transcapillary absorption. At the same time, as the interstitial fluid volume decreases, its pressure goes slightly down, thus reducing the lymph absorption and limiting the further increase of fluid absorption across the capillary wall. All these processes continue until the balance of hydrostatic and osmotic (oncotic) pressure differences is restored at the level of capillary walls, at which point the vascular refilling (lymph flow + capillary absorption) stabilises, so that the further blood volume reduction is kept at a linear limited rate.

In both Figs. 4.4 and 4.5, one can also see the initially increased capillary filtration following the saline infusion, which dilutes the blood and causes a decrease in plasma oncotic pressure. This process affects the response to dialysis over the first hour of dialysis (in other words, if following the saline infusion there was no dialysis, the cardiovascular system would reach a new steady state within roughly 1 hour).

Analogically, in the case when the priming saline is discarded, there is an initial decrease of capillary filtration due to reduced blood volume circulating within the body and the reduced capillary hydraulic pressure. Because the lymph absorption from the interstitium remains unaffected by the above, the lymph flow to the veins becomes higher than the capillary filtration leading to the refilling of the vasculature and the net increase in the blood volume with a concomitant reduction of the interstitial fluid volume. As the total blood volume increases, the capillary pressure bounces back, slowly increasing the water filtration from the capillaries and reducing the magnitude of vascular refilling until the capillary filtration is equal to the total lymph flow, at which point the system finds itself in a new steady state (again, this should take around 1 hour in case of no dialysis).

4.3 Water Shifts

The total body water is divided in the model into several compartments – extravascular extracellular water (interstitial water), extravascular intracellular water as well as plasma and red blood cell water (divided further between the individual cardiovascular compartments). In response to changes in plasma osmolarity induced by haemodialysis, the whole body becomes osmotically imbalanced, thus inducing osmotic water shifts across the cellular membranes and capillary walls. As shown in Fig. 4.6, 3.2 L of water removed during HD from the body comes mainly from the plasma and interstitial compartment, both of which are reduced in volume by slightly over 10%. Meanwhile, the cellular compartments – both red blood cells and tissue cells – loose water to a much lower extent (volume reduction of around 1%). Plasma water fraction and red blood cell water fraction are not strongly affected by haemodialysis.

Figure 4.7 shows osmotic water flows between RBCs and plasma in each of the 12 blood compartments in the model (9 cardiovascular compartments and 3 extracorporeal compartments).

As expected, the most important water shifts take place in the dialyzer and capillaries. In the dialyzer, due to diffusive loss of small solutes (especially urea and creatinine), plasma osmolarity becomes significantly reduced, which leads to the relatively high water shifts from plasma to RBCs. In the capillaries compartment, the opposite effect takes place – small solutes are absorbed from the tissues, which increases locally plasma osmolarity, and hence a water shift from RBCs to the plasma can be observed. In the remaining compartments, osmotic water shifts between RBCs and plasma are much less conspicuous, and their direction depends on the location of the given compartment with respect to the systemic capillaries and the dialyzer.

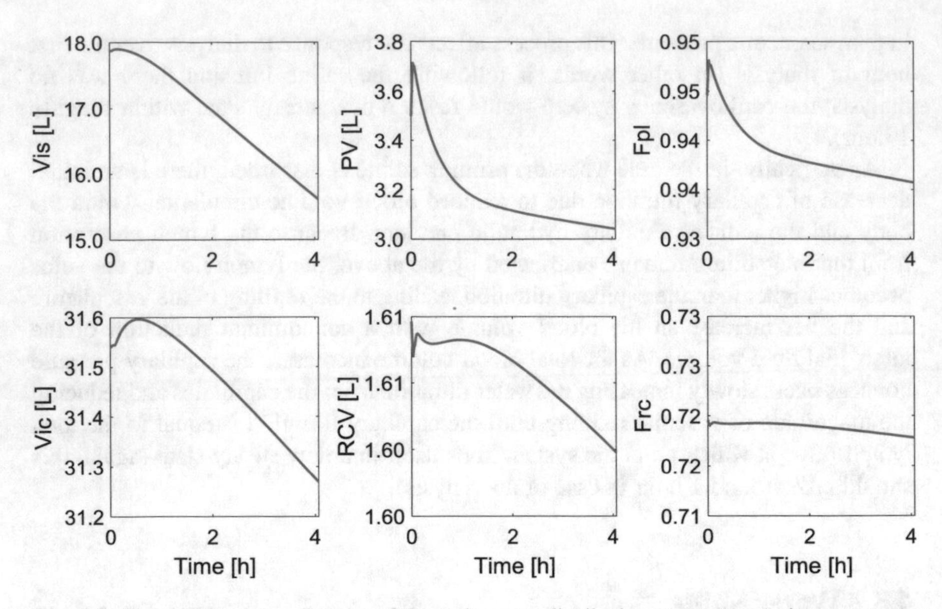

Fig. 4.6 Simulated volume changes of the main water distribution compartments in response to removal of 3 L of water from blood plasma during haemodialysis – simulations for the reference patient. *V*is, interstitial volume; *V*ic, extravascular intracellular volume (tissue cells volume); PV, plasma volume; RCV, red blood cells volume; *F*pl, plasma water fraction; *F*rc, RBC water fraction

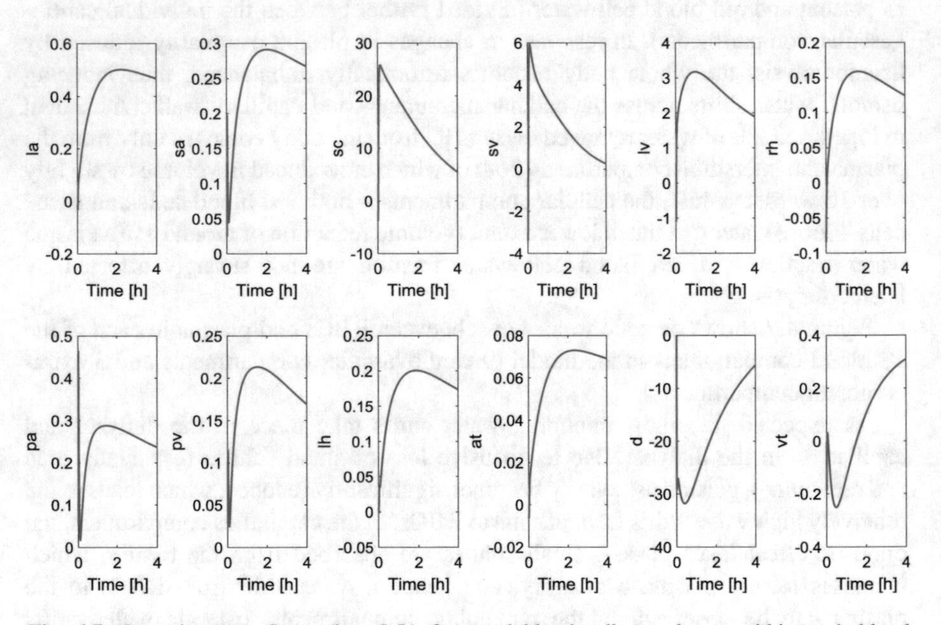

Fig. 4.7 Osmotic water flows (in mL/h) from red blood cells to plasma within each blood compartment during HD. la, large arteries; sa, small arteries; sc, systemic capillaries; sv, small veins; lv, large veins; rh, right heart; pa, pulmonary arteries; pv, pulmonary veins; lh, left heart; at, arterial tubing; d, dialyzer; vt, venous tubing

4.4 Solute Kinetics

Figure 4.8 shows changes in arterial plasma solute concentrations during the simulated haemodialysis (Case 1: priming saline infused into the patient). With the assumed dialysate fluid composition, one can observe the desired changes of individual solute concentrations in blood plasma – a reduction of high K^+ concentration, an increase of low $HCO3^-$ concentration and, of course, a significant reduction in the concentrations of urea and creatinine – the standard markers of dialysis adequacy. The concentrations of Na^+ and other cations (Ca^{2+}, Mg^{2+}) remain relatively unchanged during dialysis due to similar concentrations in plasma water and dialysate fluid (accounting for the Gibbs-Donnan effect).

At the beginning of the procedure, one can observe the effect of saline infusion, which dilutes the blood and reduces the concentrations of all solutes (especially proteins), except for the increase in the concentration of Na^+ and Cl^-, due to the higher concentration of these solutes in the saline. The concentration of other anions increases as a result of the assumed electroneutrality conditions (with the increased capillary filtration following saline infusion, the increased leakage of the negatively charged proteins is compensated by the opposite flows of other anions).

The concentration of plasma proteins (albumin and globulins) increases during haemodialysis due to both plasma volume reduction and increased amount of proteins within the vascular space (given the reduced transcapillary protein leakage, as shown in Sect. 4.2). The direction and magnitude of all solute concentration changes correspond well to the data seen in dialysis patients [8]. Altogether, the effective plasma osmolarity decreases during dialysis in the reference patient by approximately 6% (from 307 mOsm/L to 288 mOsm/L).

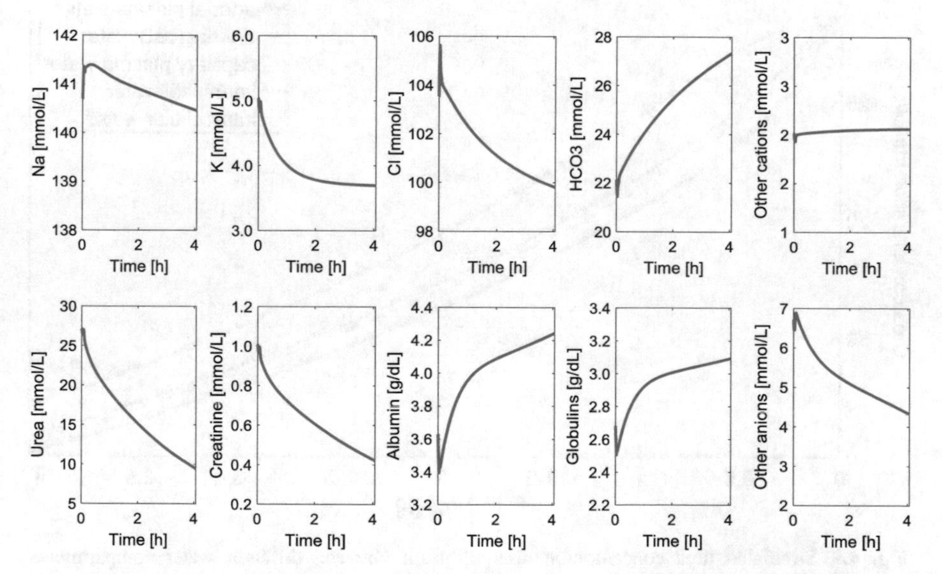

Fig. 4.8 Simulated changes of arterial plasma solute concentrations during HD procedure (Case 1: priming saline infused into the patient)

The urea reduction rate (URR) was calculated as 65% using the following formula [9, 10]:

$$\mathrm{URR}\ (\%) = \left(\frac{U_0 - U_{4\mathrm{h}}}{U_0}\right) \times 100 \tag{4.1}$$

where U_0 and $U_{4\mathrm{h}}$ are urea concentrations in arterial plasma at the beginning and at the end of dialysis, respectively.

Thusly obtained URR value is slightly lower than URR calculated from the standard single pool Kt/V (68.7%) [9, 10]:

$$\frac{Kt}{V} = -\ln\left(1 - \mathrm{URR}_{\mathrm{sp}}\right) \tag{4.2}$$

where K is urea dialyzer clearance, t is time of dialysis and V is total volume of urea distribution in the body (assumed equal to the total body water).

This discrepancy can be explained by the fact that the proposed model is a multi-pool model with a delayed transfer of urea from tissue cells to plasma and the assumed urea generation (neglected in Eq. (4.2)). A concentration gradient developing between the tissue cells and the interstitial fluid (across the cellular membrane) can be seen in Fig. 4.9. The difference in urea concentration between the interstitial fluid and capillary plasma water is relatively small, indicating a relatively unobstructed flow of urea across the capillary wall. Note the permanently higher concentration of urea in capillary plasma water (where urea is absorbed from the

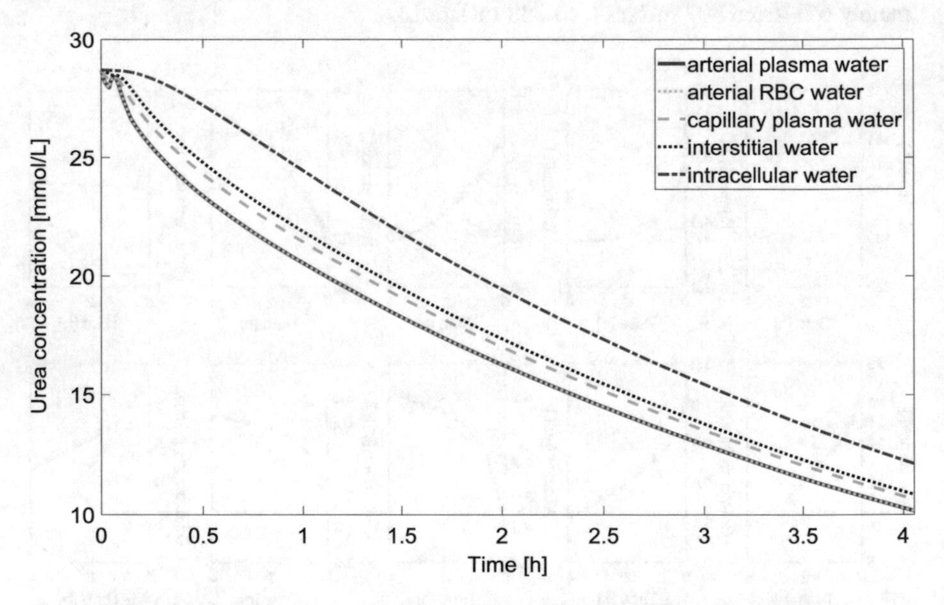

Fig. 4.9 Simulated urea concentration disequilibrium between different water compartments during the HD procedure (Case 1: priming saline infused into the patient)

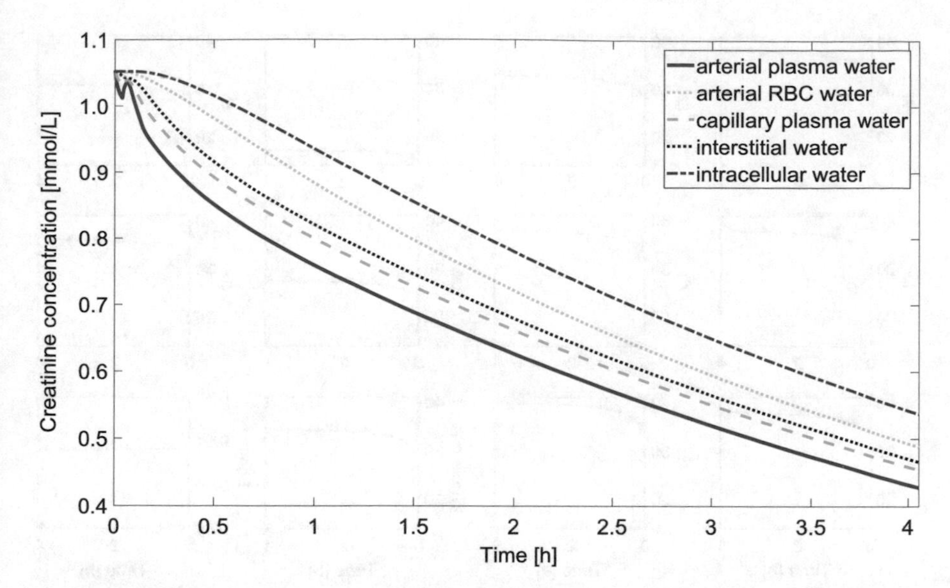

Fig. 4.10 Simulated creatinine concentration disequilibrium between different water compartments during the HD procedure (Case 1: priming saline infused into the patient)

tissues) compared to the arterial plasma water representing a mixture of urea-rich blood from the tissues, access blood bypassing the tissues and urea-depleted blood coming from the dialyzer.

With a peripheral dialysis access, as in the proposed model, the blood feeding the dialyzer is always partly diluted compared to blood leaving the tissues, which represents the so-called cardiopulmonary recirculation of dialysed blood [11]. Note also that in the arterial blood, the urea concentration is almost the same in plasma water and RBC water indicating that urea is easily transferred across the RBC membrane. The same cannot be said about creatinine, for which a gradient develops also between plasma and RBCs, as shown in Fig. 4.10.

4.5 Haematocrit Changes

As described in Sect. 2.2.5, the proposed model takes into account different levels of haematocrit across the cardiovascular system (in particular – a lower HCT in the microcirculation, compared to the macrocirculation [12–14]). During HD, because of the reduction of plasma volume, HCT increases both globally and in all individual blood compartments (see Fig. 4.11). The ratio between the whole body HCT and the central HCT (the F-cells ratio [15]) increases by circa 1%, mostly at the beginning of the procedure following the dilution of blood with the priming saline (see Fig. 4.11). During dialysis itself the F-cells ratio remains relatively stable, which is in agreement with the literature data [15, 16].

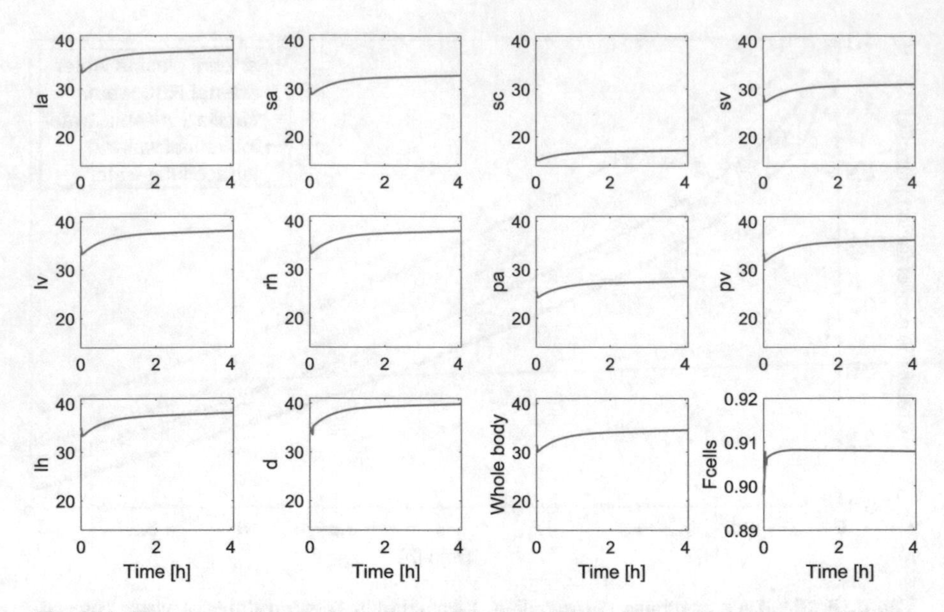

Fig. 4.11 Haematocrit changes during haemodialysis in individual blood compartments and for the whole body – model simulations for the reference patient. la, large arteries; sa, small arteries; sc, systemic capillaries; sv, small veins; lv, large veins; rh, right heart; pa, pulmonary arteries; pv, pulmonary veins; lh, left heart; d, dialyzer

In relative terms, the whole-body HCT increases by 9% (from 31.6% to 34.5%). The relative haematocrit increase is smaller in the capillaries (see Fig. 4.12), where water absorption from the tissues takes place. In the dialyzer the overall relative HCT changes are substantially higher due to the fact that the blood flowing in the dialyzer compartment is subject to the relatively higher reduction in plasma volume (a sudden increase in HCT can be observed in the dialyzer at the beginning of dialysis, when ultrafiltration is started).

Note that both Figs. 4.11 and 4.12a show the haematocrit changes counted from the beginning of the whole HD procedure, i.e. from the start of filling of the extracorporeal circuit with the patient's blood with the clearly visible effect of blood dilution by the priming saline.

Figure 4.12b shows the same simulation as Fig. 4.12a, but with the haematocrit changes counted from the beginning of the actual dialysis (i.e. circa 4 minutes after the beginning of the whole procedure). In both cases one can see the relatively sharp increase of HCT in the blood flowing through the dialyzer at the beginning of dialysis, being the effect of starting the ultrafiltration. When haematocrit changes are observed from the beginning of the procedure, there is a relatively sudden decrease of HCT in all compartments which reflects the dilution of blood by the priming saline. When HCT changes are counted from the beginning of dialysis, they are accordingly higher (+13% change of global HCT, which includes blood within the body and in the dialyzer circuit). The exact moment of starting the haematocrit measurements is hence quite important.

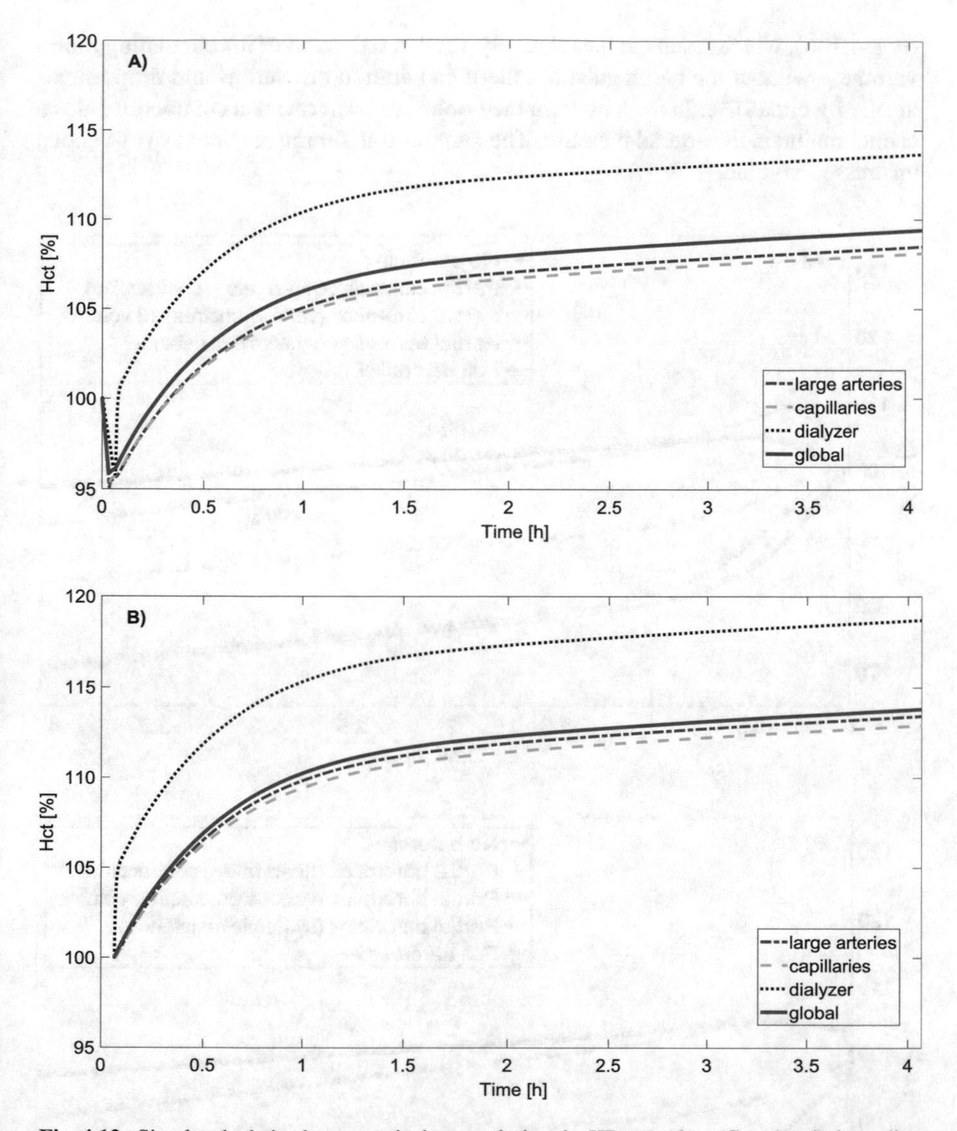

Fig. 4.12 Simulated relative haematocrit changes during the HD procedure (Case 1: priming saline infused into the patient). (**A**) Measured from the beginning of the procedure, i.e. at the start of filling the dialyzer circuit with blood. (**B**) Measured from the beginning of dialysis, i.e. 2 minutes after the dialyzer circuit had been filled with the patient's blood. Global denotes the average haematocrit of the total patient's blood including the blood in the extracorporeal circuit. All curves start at 100%

4.6 Impact of Baroreflex and Cardiac Parameters

Figure 4.13 shows the impact of individual baroreflex mechanisms on mean arterial blood pressure during the whole HD procedure. The importance of baroreflex mechanisms is best seen in the case when the priming saline is discarded

(Fig. 4.13b), which results in the relatively sudden reduction of the circulating blood volume – without the baroregulation, the mean arterial pressure would drop immediately by circa 10%. In the long term the control of heart rate and contractility alone cannot maintain the arterial pressure. The same is true for the pure control of venous unstressed volume.

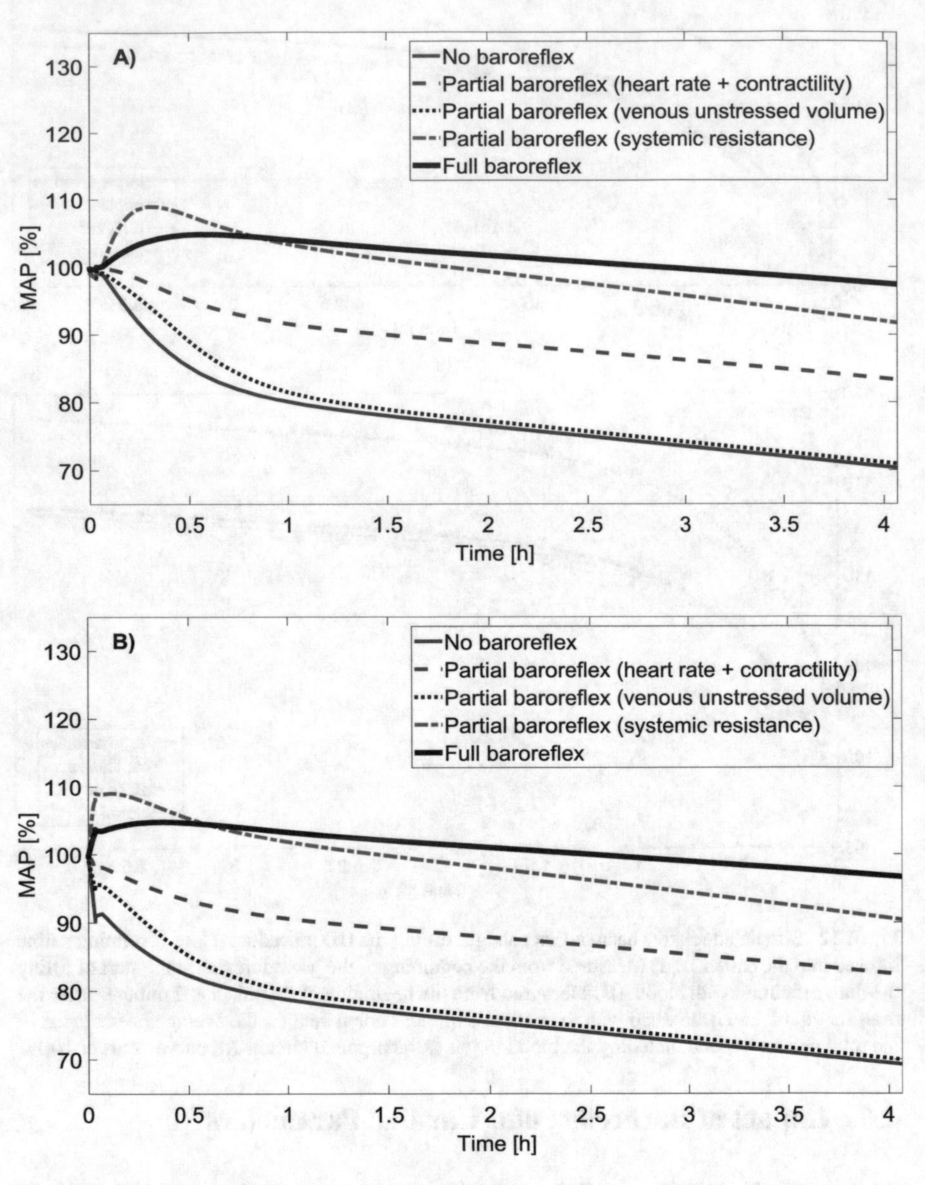

Fig. 4.13 The impact of individual baroreflex mechanisms on mean arterial blood pressure during the HD procedure. (**a**) Case 1: the priming saline infused into the patient. (**b**) Case 2: the priming saline discarded. All curves start at 100%

The arterial pressure is influenced mainly by the regulation of systemic resistance (the relatively high impact of this baroreflex mechanism was also clearly seen in our previous model of the Valsalva manaoevre [17]). However, the resistance control alone would cause a relatively large pressure overshoot following initial blood volume reduction and in the long term could be insufficient for maintaining the arterial pressure. Only the concomitant action of all four baroreflex mechanisms is able to keep the arterial pressure relatively stable throughout the simulated dialysis session. In the case when the priming saline is infused into the patient (Fig. 4.13a), the influence of individual baroreflex mechanisms is similar, with lower and more smooth changes in arterial pressure at the beginning of the procedure.

Figure 4.14 shows the impact of changing the amplitude of individual baroreflex mechanisms controlling heart period (T), vascular resistance (R), venous unstressed volume (V) and heart contractility (E). Again, it can be seen that the amplitude of the vascular resistance regulation has the highest impact on mean arterial pressure. A reduced capacity of this mechanism or the inability of the body to effectively control the vascular resistance can be clearly the source of intradialytic hypotension. On the other hand, a hyperefficient vascular resistance regulation (increased amplitude) could be the source of intradialytic hypertension with the mean arterial pressure staying above the baseline level throughout the dialysis session.

The amplitudes of other baroreflex mechanisms have a much lower impact on maintaining MAP, provided that the vascular resistance regulation is intact.

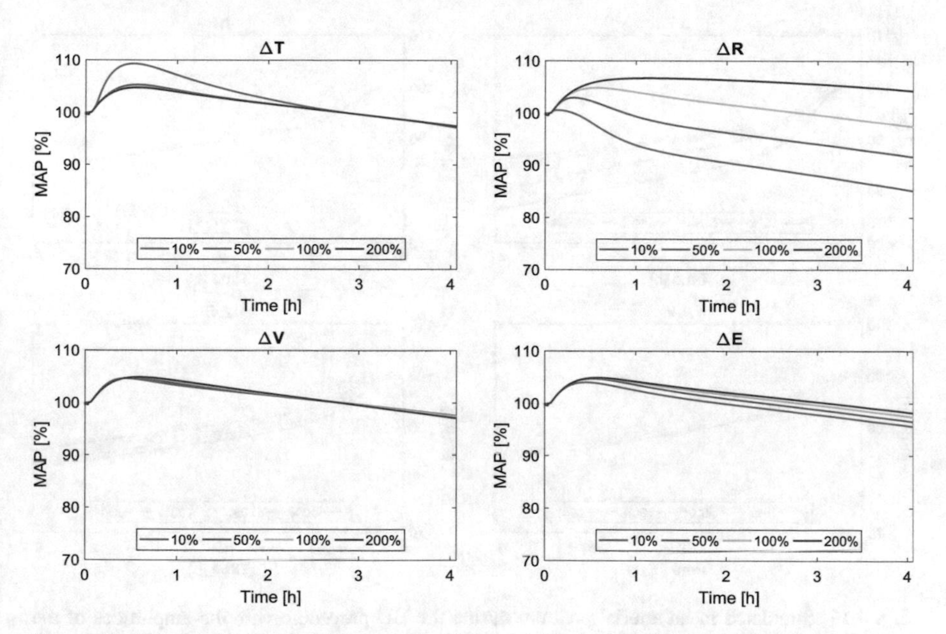

Fig. 4.14 Simulated mean arterial pressure during the HD procedure with the amplitudes of the baroreflex mechanisms controlling heart period (T), vascular resistance (R), venous unstressed volume (V) and heart contractility (E) set at 0%, 50%, 100% or 200% of the value used in the original model

However, if the mechanism controlling vascular resistance is impaired (amplitude set to 0%), the amplitudes of the remaining mechanisms become slightly more important (see Fig. 4.15).

As far as the baroreflex gains are concerned, the cardiopulmonary gain of the mechanism controlling vascular resistance has the highest impact on MAP (see Fig. 4.16). The gains of other baroreflex mechanisms seem to have a much lower influence on the long-term maintenance of a stable arterial pressure.

However, the operation of individual baroreflex mechanisms affects not only MAP but also several other cardiovascular variables. Figure 4.17 shows, for instance, how the total blood volume changes during HD are affected by different levels of the cardiopulmonary gains of the mechanisms controlling the vascular resistance and venous unstressed volume.

More complex behaviour of the cardiovascular system can be expected when multiple baroreflex parameters are affected at the same time. For instance, a decrease in some of the baroreflex gains (due to neurological impairments) could be potentially accompanied by the concomitant compensatory increase of the gains or amplitudes of other baroreflex mechanisms, thus providing a yet another pattern of cardiovascular response to haemodialysis (some of such concomitant changes in different baroreflex parameters could be seen during fitting the model to patient data in Sect. 3.2).

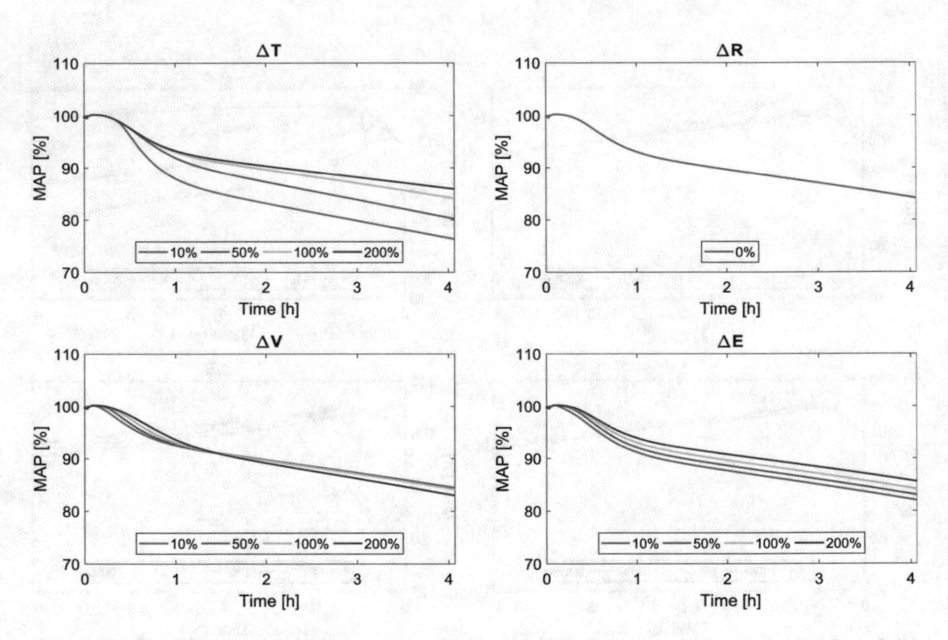

Fig. 4.15 Simulated mean arterial pressure during the HD procedure with the amplitudes of the baroreflex mechanisms controlling heart period (T), venous unstressed volume (V) and heart contractility (E) set at 10%, 50%, 100% or 200% of the value used in the original model with the amplitude of the mechanism controlling vascular resistance (R) set at 0 in all cases

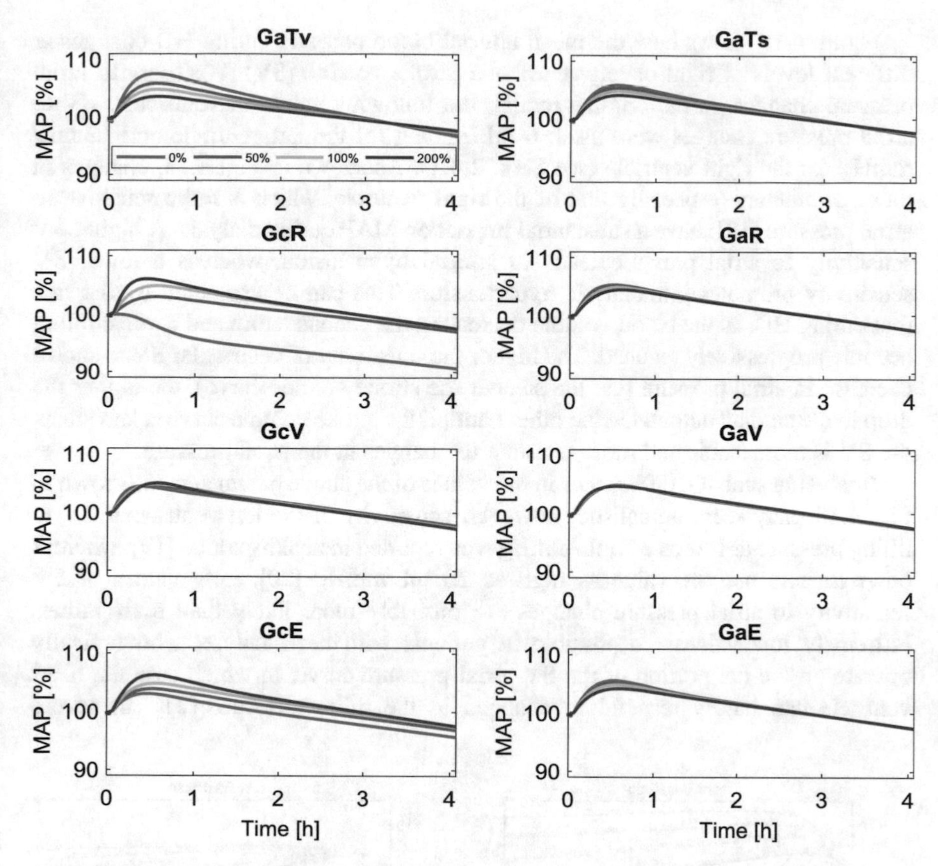

Fig. 4.16 Simulated mean arterial pressure during the HD procedure with the baroreflex gains set at 0%, 50%, 100% or 200% of the value used in the original model (in each case only one gain is changed). The legend (shown in the first panel) is the same for all panels

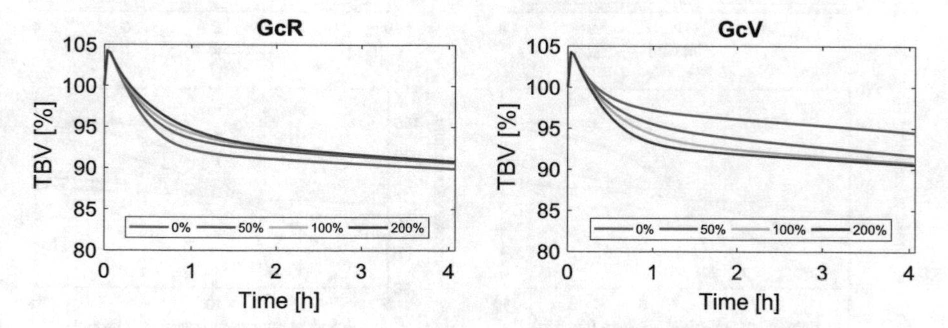

Fig. 4.17 Simulated total blood volume changes during the HD procedure with the cardiopulmonary gains of the baroreflex mechanisms controlling vascular resistance (GcR) and venous unstressed volume (GcV) set at 0%, 50%, 100% or 200% of the value used in the original model

Figure 4.18 shows how the mean arterial blood pressure during HD changes at different levels of right or left ventricular stroke volume (SV) sensitivity to atrial pressure changes. In the original model, the following values of ventricular SV to atrial pressure changes were used: 6 mL/mmHg for the left ventricle and 12 mL/mmHg for the right ventricle (see Sect. 2.4.1). As shown in Fig. 4.18, changes in these parameters (especially that of the right ventricle, which is more sensitive to atrial pressure [18]) have a substantial impact on MAP during dialysis. A higher SV sensitivity to atrial pressure leads to arterial hypotension, whereas a lower SV sensitivity promotes intradialytic hypertension. This can be explained by the fact that during HD, as the blood volume decreases, the venous return and cardiac filling become progressively reduced. The higher the sensitivity of ventricular SV to such a decrease in atrial pressure (i.e. the steeper the stroke volume curve), the higher the drop in the cardiac output. On the other hand, if the stroke volume curve is less steep, the SV is more stable and more resistant to changes in the atrial pressure.

Analysing such big differences in the values of the above parameters, as shown in Fig. 4.18, may seem unrealistic; however, sensitivity of the left ventricular SV to filling pressure as low as 2.7 mL/mmHg was reported in healthy adults [19], whereas other authors use the value as high as 20 mL/mmHg [20]. Low values of SV sensitivity to atrial pressure changes are probably more likely than high values. Extremely low values are possible in patients with heart failure, who typically operate on the flat portion of the SV-atrial pressure curve, in which case the heart ventricle can hardly respond to changes in the filling pressure [21]. In severe

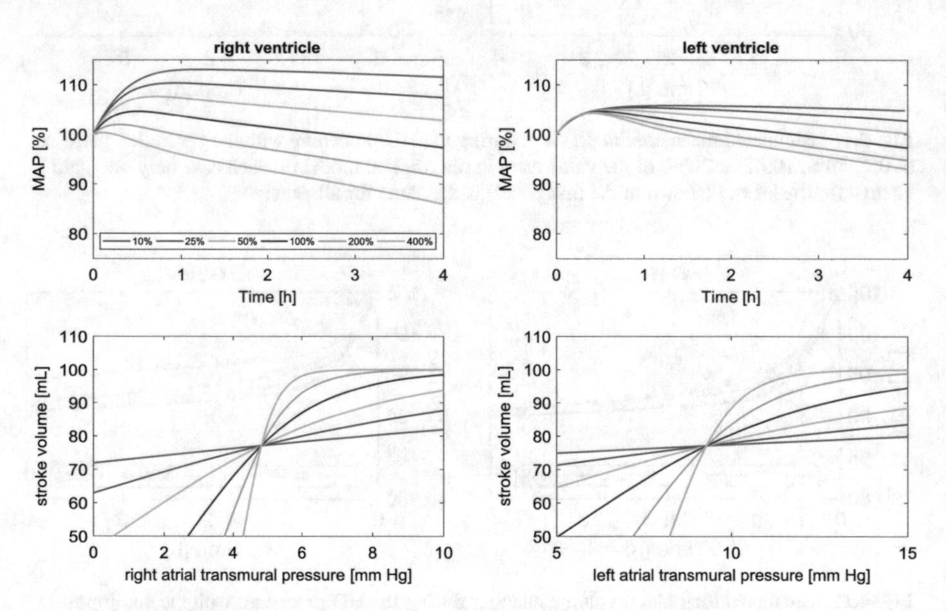

Fig. 4.18 Simulated mean arterial pressure (upper panels) and stroke volume (lower panels) of right and left ventricle during HD with the sensitivity of ventricular stroke volume to atrial pressure set at 10%, 25%, 50%, 100%, 200% or 400% of the value used in the original model. The legend (shown in the first panel) is the same for all panels

decompensated heart failure, when the heart is operating on the descending limb of the cardiac output-atrial pressure curve, the sensitivity of SV to atrial pressure changes becomes negative, so that a reduction in atrial pressure causes an increase in SV [22].

4.7 Impact of Dialysis Settings

Figure 4.19 shows the simulated mean arterial pressure changes during HD at different levels of Na^+ and Cl^- concentration in the dialysate fluid (within the range seen in the clinical practice). One can see that, even though different levels of these ions in the dialysate fluid affect the changes in their plasma concentrations during HD (right panels in Fig. 4.19), their impact on mean arterial pressure is rather limited, especially in the case of chloride. For other ions analysed in the model (i.e. potassium, bicarbonate and other cations), the analogical changes in MAP are almost negligible (which is expected given their lower plasma levels), and hence the similar graphs are not shown for those ions.

Assuming that the patient has a normal level of plasma Na^+ and Cl^-, the dialysate levels of these ions should be close to the levels seen in the patient's plasma corrected for plasma water fraction and the Gibbs-Donnan coefficient. For sodium, given that the plasma water fraction and the Gibbs-Donnan coefficient are almost similar (~0.94) and cancel out each other, these corrections can be practically

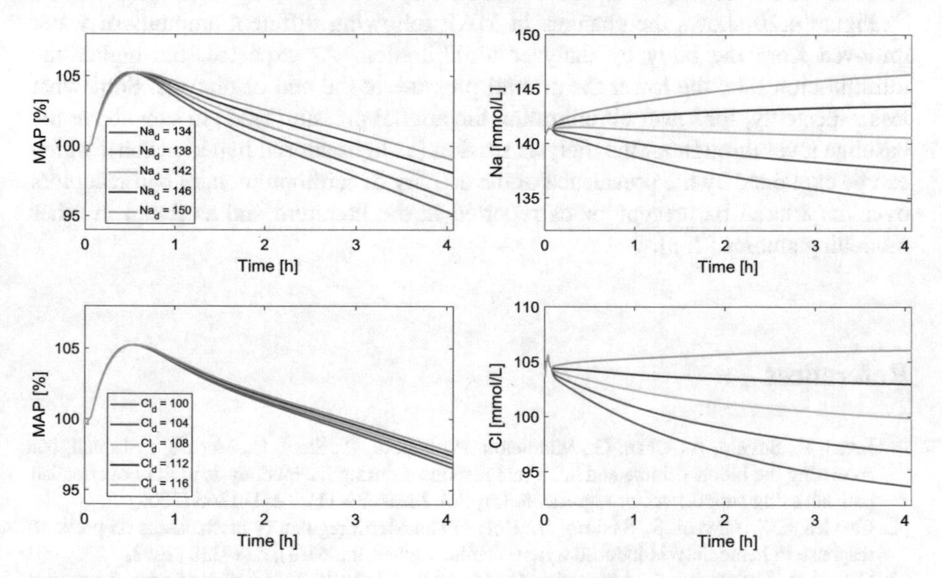

Fig. 4.19 Simulated mean arterial pressure during HD for different levels of sodium (Na^+) or chloride (Cl^-) concentration in the dialysate fluid (in mmol/L). The panels on the right-hand side show the corresponding patterns of plasma concentrations of Na^+ and Cl^- for different levels of Na^+ and Cl^- in the dialysate fluid (the legends from the left panels apply)

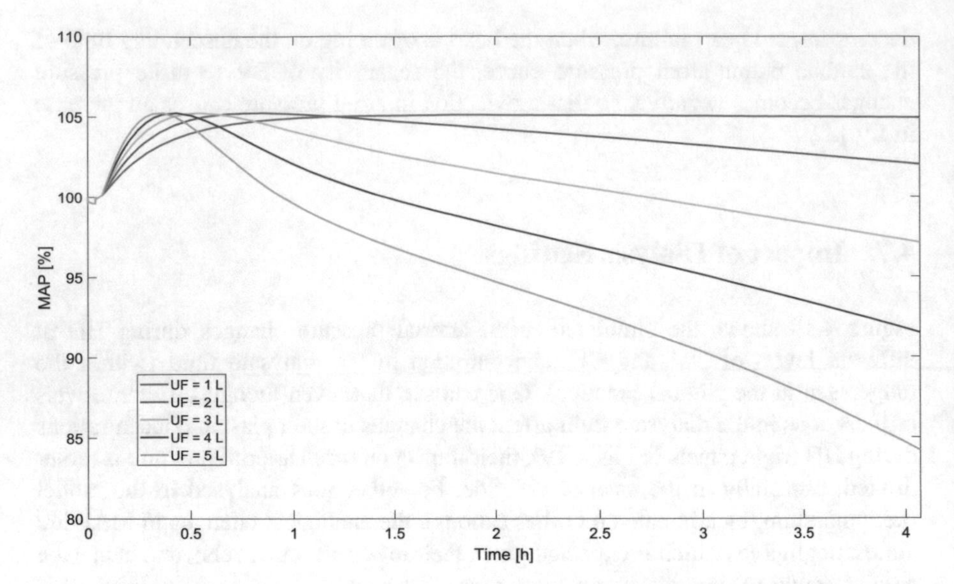

Fig. 4.20 Simulated mean arterial pressure during HD for different total dialyzer ultrafiltration

ignored; however, for chloride the Gibbs-Donnan coefficient is slightly above 1 (the empirically found value 1.01 was assumed in this study [23, 24]), and hence the dialysate level of chloride should be appropriately higher than its plasma level in order not to reduce its plasma concentration below physiologically acceptable level.

Figure 4.20 shows the changes in MAP following different amounts of water removed from the body by dialyzer ultrafiltration. As expected, the higher the ultrafiltration rate, the lower the arterial pressure at the end of dialysis. Somewhat less expectedly, for lower ultrafiltration the arterial pressure tends to stay above the baseline level throughout the dialysis session (a slight arterial hypertension), which can be explained by the prevalence of the activity of cardiopulmonary baroreceptors over the arterial baroreceptors, as reported in the literature and as found in other modelling studies [2, 3].

References

1. Javed, F., Savkin, A., Chan, G., Middleton, P., Malouf, P., Steel, E., Mackie, J., Lovell, N.: Assessing the blood volume and heart rate responses during haemodialysis in fluid overloaded patients using support vector regression. Physiol. Meas. **30**(11), 1251–1266 (2009)
2. Cavalcanti, S., Cavani, S., Santoro, A.: Role of short-term regulatory mechanisms on pressure response to hemodialysis-induced hypovolemia. Kidney Int. **61**(1), 228–238 (2002)
3. Cavani, S., Cavalcanti, S., Avanzolini, G.: Model based sensitivity analysis of arterial pressure response to hemodialysis induced hypovolemia. ASAIO J. **47**(4), 377–388 (2001)
4. Pietribiasi, M., Katzarski, K., Galach, M., Stachowska-Pietka, J., Schneditz, D., Lindholm, B., Waniewski, J.: Kinetics of plasma refilling during hemodialysis sessions with different initial fluid status. ASAIO J. **61**(3), 350–356 (2015)

5. Pietribiasi, M., Wojcik-Zaluska, A., Zaluska, W., Waniewski, J.: Does the plasma refilling coefficient change during hemodialysis sessions? Int. J. Artif. Organs. **41**(11), 706–713 (2018)
6. Levick, J.: Capillary filtration-absorption balance reconsidered in light of dynamic extravascular factors. Exp. Physiol. **76**(6), 825–857 (1991)
7. Levick, J., Michel, C.: Microvascular fluid exchange and the revised Starling principle. Cardiovasc. Res. **87**, 198–210 (2010)
8. Kyriazis, J., Kalogeropoulou, K., Bilirakis, L., Smirnioudis, N., Pikounis, V., Stamatiadis, D., Liolia, E.: Dialysate magnesium level and blood pressure. Kidney Int. **66**(3), 1221–1231 (2004)
9. Daugirdas, J., Blake, G., Ing, T. (eds.): Handbook of Dialysis, 5th edn. Wolters Kluwer Health, Philadelphia (2015)
10. Debowska, M., Lindholm, B., Waniewski, J.: Kinetic modeling and adequacy of Dialysis. In: Carpi, A. (ed.) Progress in Hemodialysis – From Emergent Biotechnology to Clinical Practice. InTech (2011)
11. Schneditz, D., Kaufman, A., Polaschegg, H., Levin, N., Daugirdas, J.: Cardiopulmonary recirculation during hemodialysis. Kidney Int. **42**(6), 1450–1456 (1992)
12. Cokelet, G.: Speculation on a cause of low vessel hematocrits in the micro circulation. Microcirculation. **2**, 1–18 (1982)
13. Fåhraeus, R.: The suspension stability of blood. Physiol. Rev. **9**, 241–274 (1929)
14. Fournier, R.: The physical and flow properties of blood. In: Basic Transport Phenomena in Biomedical Engineering, 3rd edn. CRC Press, Boca Raton (2011)
15. Schneditz, D., Ribitsch, W., Schilcher, G., Uhlmann, M., Chait, Y., Stadlbauer, V.: Concordance of absolute and relative plasma volume changes and stability of Fcells in routine hemodialysis. Hemodial. Int. **20**, 120–128 (2016)
16. Chaplin Jr., H., Mollison, P., Vetter, H.: The body/venous hematocrit ratio: its constancy over a wide hematocrit range. J. Clin. Invest. **32**(12), 1309–1316 (1953)
17. Pstras, L., Thomaseth, K., Waniewski, J., Balzani, I., Bellavere, F.: Mathematical modelling of cardiovascular response to the Valsalva manoeuvre. Math. Med. Biol. **34**(2), 261–292 (2017)
18. Rothe, C.: Reflex control of veins and vascular capacitance. Physiol. Rev. **63**(4), 1281–1342 (1983)
19. Levine, B., Lane, L., Buckey, J., Friedman, D., Blomqvist, C.: Left ventricular pressure-volume and Frank-Starling relations in endurance athletes. Implications for orthostatic tolerance and exercise performance. Circulation. **84**(3), 1016–1023 (1991)
20. Ursino, M., Antonucci, M., Belardinelli, E.: Role of active changes in venous capacity by the carotid baroreflex: analysis with a mathematical model. Am. J. Physiol. Heart Circ. Physiol. **267** (6), H2531–H2546 (1994)
21. Guyton, A., Hall, J.: Textbook of Medical Physiology, 11th edn. Elsevier Saunders, Philadelphia (2006)
22. Sarnoff, S., Berglund, E.: Ventricular function. I. Starling's law of the heart studied by means of simultaneous right and left ventricular function curves in the dog. Circulation. **9**(5), 706–718 (1954)
23. Manery, J.: Water and electrolyte metabolism. Physiol. Rev. **34**(2), 334–417 (1954)
24. Swan, R., Feinstein, H., Madisso, H., Plinke, A., Sharma, H.: Distribution of sulfate ion across semi-permeable membranes. J Clin Invest. **35**, 607–610 (1956)

Chapter 5
Conclusions, Challenges and Directions for Future Research in Haemodialysis Modelling

Abstract This chapter summarises the presented mathematical model of the cardiovascular system, body fluid shifts and solute kinetics during haemodialysis with a brief overview of the performed sensitivity analysis and model validation as well as the main results of model simulations. The authors share also challenges they faced during the work on the model and indicate the possible directions for future research with the proposed model.

Keywords Model scope · Simulation results · Challenges · Parameter values · Initial conditions · Long-term blood pressure regulation · Model complexity · Clinical data acquisition · Future research · Haemodialysis optimization · Dialysis efficacy · Sodium modelling · Ultrafiltration profiling · Global sensitivity analysis · Regional blood flow · Fluid infusions

5.1 Summary

5.1.1 Model Scope

This book presents an integrated compartmental model of the whole-body distribution and transport of blood, water, proteins (albumin and globulins), electrolytes (sodium, potassium, chloride, and bicarbonate ions) as well as electrically neutral solutes (urea and creatinine). Thanks to the full integration of the baroreflex-equipped model of the cardiovascular system with the description of the whole-body transport of water and solutes, the proposed model is able to simulate changes in volume, pressure and solutes concentrations in all considered fluid compartments (plasma, red blood cells, interstitium and tissue cells), and hence it can provide insights into the processes taking place across the human body, both under normal conditions and during haemodialysis, from the mathematical model perspective. The model is able to simulate the whole HD procedure including filling the extracorporeal circuit with the patient's blood. It can be used for a thorough analysis

L. Pstras, J. Waniewski, *Mathematical Modelling of Haemodialysis*,
https://doi.org/10.1007/978-3-030-21410-4_5

of the human's body response to haemodialysis for different conditions of dialysis treatment or for different pre-dialysis states of the patient.

5.1.2 Model Validation

The model has been validated by fitting it to clinical data from dialysed patients from the Lublin Medical University, Poland (intradialytic data on mean arterial blood pressure, morphologic blood parameters, such as HCT, HGB, MCV, RBC and arterial plasma concentrations of main analysed solutes – sodium, potassium, urea, creatinine, albumin, total protein), as well as to clinical data available from the literature (haemodynamic parameters only). In both cases, following tuning of selected model parameters with a clear physiological meaning, the model was able to fit solute and/or haemodynamic data with high accuracy (the average root mean squared error of circa 1.5%).

5.1.3 Sensitivity Analysis

The presented model turned out to be relatively robust and resistant to small (infinitesimal) changes in all model parameters during a standard 4-hour HD session with only a limited number of parameters to which the relative sensitivity of arterial blood pressure (MAP) was above 0.1 meaning that a change in the parameter value by 10% will affect the studied model outcome by less than 1% (all relative sensitivities were below unity). The local sensitivity analysis of MAP presented in Sect. 3.1 identified the key model parameters, most of which are related to the transport of proteins, the vascular refilling mechanism and the exchange of sodium and chloride across the dialyzer membrane.

5.1.4 Simulation Results

The haemodynamic response to HD has been simulated for the reference anuric patient with different dialysis settings (different dialyzer ultrafiltration rates and different composition of the dialysate fluid). Through model-based simulations, it was shown that with well-functioning cardiovascular regulatory mechanisms and an efficient vascular refilling mechanism, the patient should tolerate well the removal of up to circa 4 L of water during dialysis with the mean arterial pressure decreasing by no more than circa 8%. As expected, the higher the ultrafiltration rate, the lower the arterial pressure at the end of dialysis. Somewhat less expectedly, for lower ultrafiltration rates, the arterial pressure tends to stay in the simulations above the baseline level throughout the dialysis session, which can be explained by the prevalence of the activity of cardiopulmonary baroreceptors over the arterial baroreceptors according to the assumed values of the baroreflex parameters [1, 2]. The concentration of individual ions in the dialysate fluid (within the typically used range) has a relatively small impact on the blood pressure response to dialysis, with sodium having the highest, albeit still small, influence on MAP.

Two cases were considered and compared in the book: the case when the fluid filling the dialyzer circuit before the session is infused into the patient and the case when the priming fluid is discarded, thus reducing the total circulating blood volume. Both cases feature a similar ultimate response of the cardiovascular system but differ at the beginning of the procedure, thus indicating that the haemodynamic changes during HD are induced not only by the dialysis treatment itself but also reflect the prior drawing of some of the patient's blood to fill the extracorporeal circuit. In particular it was shown, how the regulatory mechanisms respond to the sudden "loss" of blood volume circulating within the body, when the priming fluid is disposed of to the drain bag.

On top of the local sensitivity analysis, a separate analysis was provided to determine the impact of significant changes in the baroreflex parameters and cardiac stroke volume sensitivity to atrial pressure changes and hence to analyse how the potential impairments or overactivity of the above mechanisms affect the arterial pressure response to HD. It was shown that the parameters of the mechanism controlling the vascular resistance (especially the amplitude and the cardiopulmonary gain) have a substantially higher impact on MAP during haemodialysis compared to other baroreflex mechanisms. A reduced capacity of this mechanism can be clearly the source of intradialytic hypotension. On the other hand, a hyperefficient vascular resistance regulation (its increased amplitude) could be the source of intradialytic hypertension with the mean arterial pressure staying above the baseline level throughout the dialysis session. It was also shown that the sensitivity of ventricular stroke volume to atrial pressure (especially that of right ventricle) has an important impact on MAP during dialysis, with a high cardiac sensitivity leading to arterial hypotension and a low sensitivity promoting intradialytic hypertension.

The model and the presented model-based simulations constitute a certain step towards a better understanding of the complex behaviour of the cardiovascular system in response to dialysis-induced changes in blood volume and plasma osmolarity; however, further simulation studies and analyses based on extensive clinical datasets are still needed. The proposed model provides a solid framework for such future modelling efforts.

5.2 Challenges

Developing complex mathematical models, such as the one presented in this book, is often associated with various challenges related to the process of defining the structure of the model, making necessary assumptions, assigning the values of model parameters or performing numerical computations. Below are discussed some of such challenges associated with the present model, which are likely to be faced in other, similar models.

5.2.1 Parameter Values

As indicated in Chap. 2, the values of all model parameters were taken from the literature for the reference healthy subject. For many parameters, however, different sources often provide different values (sometimes largely different). In such cases, the choice of parameter values for our model was always based on a compromise between available data in an attempt to provide the best possible description of an average subject. On the other hand, in several cases the data on normal parameter values was either difficult to find (e.g. permeability of red cell membrane to bicarbonate ions) or even impossible to find (e.g. reflection coefficient of urea at the tissue cell membrane or reflection coefficient of creatinine at the red blood cell membrane). For the latter, the parameter values were assumed based on the analogous values for similar parameters (e.g. for other types of cells or other solutes). Yet in other cases, the data available in the literature was not directly transferrable to our model given its structure (i.e. the assumed compartmentalisation). For instance, we did not manage to find directly data on normal compliance of all individual vascular compartments as present in the model. Instead, we had to use data on total human vascular compliance and compliance of certain parts of the circulatory system and then divide it according to our model structure based on certain assumptions or auxiliary data from other sources, as discussed in Sect. 2.4.1. As a consequence, even though we tried our best to describe an average subject in the best possible way, given the shortage or lack of some data and the number of different literature sources used to obtain those data, our virtual reference subject may not be an ideal representation of an average person.

5.2.2 Initial Conditions

Assigning initial conditions of the modelled system is even more problematic than assigning the values of model parameters. This is particularly the case for the initial pressures and volumes of all cardiovascular compartments or the initial volumes of extravascular compartments, i.e. the extravascular extracellular compartment (the interstitial fluid) and the extravascular intracellular compartment. Even if we managed to describe properly the initial conditions for an average healthy subject (based on literature data on typical pressures and volumes of individual vascular compartments, typical distribution of total body water and typical water fractions of extravascular compartments), assigning initial conditions of the system for individual patients remains a big challenge. For instance, if one wanted to simulate a patient with an elevated arterial blood pressure (e.g. 160/100 mmHg), one can easily adjust the initial mean pressure in the arterial compartment to a given value (in this case 120 mmHg), but assigning the initial pressures in other vascular parameters remains a puzzling enigma, given that those pressures may or may not be elevated and cannot be measured non-invasively (especially on the whole-body level). Assigning the initial volumes of individual cardiovascular compartments is even more challenging, as one has no information whatsoever on the actual distribution of blood across the circulatory system in the given patient (an elevated arterial blood pressure is most likely associated with an altered distribution of blood in the system, i.e. a larger amount of blood in the arterial tree, but this may not be necessarily the case, if the total blood volume is different from expected). Note also that using data from different sources may sometimes lead to inconsistencies in describing the initial conditions of the modelled system. For instance, as indicated in Sect. 2.5.1, based on the literature data, we assigned 35.7% and 53.7% of total body water to the extravascular extracellular and extravascular intracellular fluids, respectively. This implies that the remaining 10.6% of the total body water must be contained within the vascular system (in plasma and in red blood cells). If, however, the total blood volume was estimated from one of the many available formulae (based on the anthropometric parameters), the water content of blood (assuming a certain level of haematocrit and using the widely accepted values for water fraction in plasma and in red blood cells) could constitute a different percentage of the assumed total body water (e.g. 9% or 12%). This was the reason why we decided to derive the total blood volume from the assumed water content in blood, rather than using a formula-based estimate.

5.2.3 *Patient Before Dialysis*

In the study presented in this book, we started from defining the model for the steady-state conditions in a reference healthy subject and then made a few modifications (perturbations) in the modelled system to reach a new steady state of the system representing a patient before dialysis. Those modifications included adding a certain amount of water to the system to represent fluid overload, adjusting the amounts of selected solutes in the system to represent typical plasma solute concentrations seen in dialysis patients, removing a certain amount of erythrocytes from the cardiovascular system to reduce the haematocrit level, and adding the arteriovenous fistula to the system to represent dialysis access. The main reason for this approach (i.e. defining the model for a healthy subject and then reaching the pre-dialysis state through model simulation) was to avoid the aforementioned problematic issue of assigning the initial volumes and pressures of all compartments, which would be needed if we wanted to define the system straight off for a pre-dialysis patient. However, the quid pro quo was that this approach was in fact a simulation of the long-term changes in the human body following kidney failure and surgical creation of an arteriovenous fistula, which, by significantly changing the whole-body water-electrolyte balance and blood flow conditions (reduced resistance due to fistula and a decreased level of haematocrit), normally activates a number of regulatory and compensatory mechanisms, which are not accounted for in the model (e.g. long-term neurohormonal blood pressure regulation or cardiac remodelling). Given the above, we proposed a simplified approach to reflect these long-term systemic changes by additionally modifying two model variables: the initial pre-dialysis resistance of small arteries and cardiac contractility (both being later regulated in the model by the baroreflex) in order to reach the desired pre-dialysis arterial blood pressure and cardiac output, which can be relatively easily measured or estimated in real patients. This is, however, a rather simplified approach introduced in our model, which was even further simplified by the assumption on the baroreceptors being reset around the new operating blood pressure. The actual long-term neurohormonal changes, cardiac remodelling and resetting of baroreceptors are in fact much more complex and influence many more variables or parameters than just those two adjusted in our model. A possible alternative, simplified approach to reaching the desired pre-dialysis arterial blood pressure and cardiac output could be a modification of the parameters of the left and right ventricular stroke volume-atrial pressure curves (e.g. shifting of the stroke volume-atrial pressure curve, as proposed in our previous modelling studies [3, 4]).

5.2.4 Transport Parameters

In the simulations shown in this book, all parameters of solute and water transport in the reference dialysis patient were assumed the same as for the reference healthy subject. In reality, however, the transport parameters in dialysis patients may not only differ from the analogous parameters in healthy people but may also change during dialysis due to possible alterations in the properties of the capillary walls or cellular membranes (e.g. due to cell shrinkage). Moreover, in the case of capillary walls, the direction of fluid transport is reversed during dialysis (from capillary filtration to absorption), which can also affect the transcapillary transport properties (kept in the model the same regardless of the flow direction).

5.2.5 Number of Parameters

The presented model has a relatively large number of parameters (a total of circa 150, including the values describing the initial conditions of the system). This results mainly from the relatively high number of compartments included in the model and the similarly high number of solutes analysed (all solutes are associated with separate parameters describing their transport properties). Even though all parameters used in the model have a clear physiological meaning, many of them cannot be easily and non-invasively measured in patients or cannot be measured at all (e.g. lumped resistances and compliances of individual vascular compartments, whole-body hydraulic conductivity of the capillary walls, etc.). This has two important implications for using the model for simulating individual patients. Firstly, only a very limited number of parameters can be given values measured in the patient (e.g. arterial blood pressure, heart rate, central haematocrit, plasma solute concentrations). Secondly, in case of fitting the model to clinical data, it is not easy to decide which parameters should be (or can be) estimated from the data. The sensitivity analysis should provide some guidance on this issue indicating the parameters to which the model is relatively sensitive and the parameters with little or no influence on the behaviour of a given model output or outputs. This approach was employed in the present study for fitting the model to clinical data. The sensitivity analysis showed that only a limited number of parameters have a significant impact on the blood pressure response to HD, with the majority of parameters having little or no practical effect on intradialytic haemodynamics. Therefore, we kept most of model parameters at the assumed normal physiological values, while adjusting only the parameters to which the studied model output is relatively sensitive to. Additionally, the fitting procedure was designed to adjust also the values of certain parameters reflecting some individual pathologies or presenting significant inter-subject differences (for instance, the baroreflex parameters or the parameters influencing the capillary filtration rate, which can vary significantly between different patients [5]). Note, however, that the sensitivity analysis presented in Sect. 3.1 was only a local sensitivity analysis with respect to a single model output (mean arterial blood pressure), and hence for a different study, a separate sensitivity analysis may be needed (depending on the studied model output or outputs) and could yield a different set of parameters the model is sensitive to.

5.2.6 *Model Complexity*

In the process of developing the model, we tried to achieve a balance between the level of model complexity and the accuracy of describing physiological processes or phenomena in an attempt to produce a model that would be relatively simple to use and understand while being physiologically sound and solid. In its final presented form, the model may seem relatively complex, but it is still fairly simple (and accordingly less detailed) compared to more comprehensive models of human physiology, such as the famous model of the human circulation proposed by Guyton et al. [6], later presented as the HUMAN model by Coleman and Randall [7]. Even though not all of the features implemented in our model were crucial for the presented simulations, we decided to include them in the model, given that they may become important for simulating other, more complex dialysis treatments in individual patients with different pathologies, or for other analyses and simulations not shown in this book. For instance, simulating protein transport separately for albumin and non-albumin proteins may be important for patients with abnormal albumin/globulins ratio. Similarly, having a separate compartment for arterial and venous tubing in the extracorporeal circuit may enable simulations of pre- or post-dilution haemofiltration therapy. Finally, several aspects of the cardiovascular system included in the proposed model (such as the nonlinear Frank-Starling mechanism, the relationship between the vessel volume and its resistance to blood flow or the dependence of blood viscosity on haematocrit) may not have a significant effect on the simulations of a standard, uncomplicated dialysis session, but may have a much higher significance during a more intense dialysis with higher ultrafiltration leading to a larger reduction in blood volume and cardiac output, a more significant increase of haematocrit or a possible hypotensive event.

5.2.7 *Haemodialysis Procedure*

As indicated in Sects. 2.6 and 2.7, in our simulations we analysed the whole haemodialysis procedure including filling of the extracorporeal circuit with the patient's blood. In clinical settings various types of dialyzers and dialysis machines are used (which affects not only dialysis efficiency but also the total priming volume of the extracorporeal circuit), and different dialysis units follow different scenarios of running the dialysis procedure in terms of timing of the procedure, the blood pump speed settings, infusion or removal of the priming fluid, etc. In this study we assumed a certain volume of the dialyzer and dialysis tubing (220 mL) and the blood pump speed of 100 mL/min for filling the circuit with patient's blood. Then, for modelling the beginning of the actual HD, we assumed an idealised case of a linear increase of the blood pump speed from the previously set level of 100 mL/min to the target level of 300 mL/min over the course of 2 minutes (with the dialyzer in the bypass mode, i.e. with no ultrafiltration and a negligible diffusion of solutes). Given the time scale of dialysis of a few hours, those few minutes of initiating the dialysis session may seem irrelevant, but as shown in Chap. 4, this pre-dialysis procedure has a substantial impact on the cardiovascular system, furthermore depending on whether the priming saline is infused to the patient or discarded (the initiation of HD is in general associated with a strong impact on the cardiovascular system [8, 9]). For a complete analysis of the patient's cardiovascular system from the very beginning of the dialysis session, the details of this pre-dialysis procedure are hence worth to be included in the model.

5.3 Future Research and Potential Applications of the Model

The proposed model provides a comprehensive tool for analysing the haemodynamic response to haemodialysis and investigating the interactions between the cardiovascular system and reflex regulatory mechanisms in response to blood volume and pressure reduction due to water and solute removal in the dialyzer. Following further improvements and extensions, the model could be eventually used for optimising HD treatment to enable efficient removal of waste products and excess water from the body, while maintaining patient's cardiovascular stability. In particular, it could be used for analyses of HD with variable settings (e.g. profiling of sodium in the dialysate fluid or profiling of the ultrafiltration rate) or analyses of different dialysis modalities, e.g. continuous veno-venous haemodialysis or haemofiltration used as a renal replacement therapy in acute kidney injury.

Using the model for simulating real patients should be preceded by adjusting the parameters for the individual patient data to account for physiological or pathophysiological differences between the patients. As already mentioned this applies only to those parameters, which can be measured in the patient. For example, the plasma concentrations of all analysed solutes should be adjusted in the model accordingly, as an unusually low plasma protein concentration or excessively high plasma urea concentration can both affect the transport and regulatory processes taking place during haemodialysis.

Future research using the developed model should concentrate on fitting the model to larger datasets from different patients and different dialysis settings, both for further model validation and for testing various hypotheses concerning physiological or pathophysiological mechanisms and transport processes taking place during haemodialysis. Fitting the model to data from complicated dialysis sessions featuring abnormal blood pressure responses, such as intradialytic hypo- or hypertension, could be particularly helpful for increasing the understanding of intradialytic hypotension phenomenon, which remains the major issue in dialysis units. Data from profiled dialysis sessions, i.e. dialysis with variable settings, such as variable ultrafiltration rate or variable composition of the dialysate fluid, could be particularly useful for an extensive model validation. Following such validation, the model could be potentially employed for identifying the optimal conditions of dialysis (ultrafiltration rate, dialysate fluid composition, etc.) in hypotension-prone patients in order to reduce the risk of intradialytic hypotension. Note, however, that, unlike for haemodynamic parameters, such as heart rate or blood pressure, which can be monitored during haemodialysis non-invasively, it is not always easy to obtain data requiring additional, in some cases invasive, measures, such as drawing of blood for laboratory measurements, given the usually limited availability of medical personnel and the difficulties in obtaining patient consent or bioethical committee approval (especially for elderly or fragile patients, patients with extremely low haematocrit, or hypotension-prone patients in whom the already problematic dialysis-induced blood volume reduction should not be further deepened by blood sampling). The emerging technologies for automatic, non-invasive blood analysis could help in this matter. Note also that from the modelling point of view, it would be desirable to validate the model using data from several dialysis sessions with different settings in the same patient, but obviously the dialysis settings cannot be freely adjusted against the dialysis prescription, and, besides, the patient's pre-dialysis fluid status may vary from session to session. Finally, model validation could be further improved by fitting the model not only to intradialytic data but also to data from certain time after dialysis, in particular data on concentrations of individual solutes in the blood with the well-known post-dialysis rebound effect.

To facilitate and improve the quality of fitting the model to various complex datasets, a global sensitivity analysis of the model should be performed to investigate the impact of large parameter changes of all model parameters or local changes around different set of parameter values (e.g. using Monte Carlo simulations), ideally studied with respect to different model outputs (in this book we presented only a local sensitivity analysis of one model output to small parameter changes around the assumed normal values). Such an extended global sensitivity analysis could enable a reduction of the model structure and the number of model parameters for the given task, thus reducing the computational requirements of the fitting procedure and improving the significance of the obtained results.

The model is presented in its integrated entirety, although it can be subdivided into several substructures, which can be potentially studied separately. These include (1) water and solute exchange across the cellular membrane, (2) transcapillary water and solute exchange, (3) ultrafiltration and solute exchange between plasma and dialysate fluid and (4) blood flow across the cardiovascular system (with or without individual baroreflex mechanisms). Based on appropriate clinical data, in the future these model substructures could be additionally tested and validated independently.

Future versions of the model could be extended to incorporate different properties of regional blood subcirculations (to individual organs or groups of highly perfused and low-perfused organs [10]) or cardiovascular regulatory mechanisms other than baroreflex, such as organ autoregulation (especially for heart and brain) or humoral regulation. A pulsatile version of the model could be developed to investigate changes of systolic and diastolic blood pressure during HD and to analyse the impact of pulsatile blood flow on the operation of the baroreflex mechanisms (following the appropriate changes in the baroreflex model itself). The model could be employed to analyse patient's response not only to a single dialysis session but to a longer period of haemodialysis therapy (e.g. to a week dialysis cycle), which would, however, require adding to the model long-term blood pressure regulatory mechanisms. The model of transcapillary transport could be extended to include different types of pathways for solute and water transport through the capillary walls (e.g. pores of different sizes as in the two-pore or three-pore theory). The model could be also used for the assessment of the impact of access or cardiopulmonary recirculation on the dialysis efficacy (if needed, the model can be modified to include a central veno-venous access instead of the arteriovenous fistula). Finally, a variable parameter version of the model could be studied to investigate the impact of potential changes of certain parameters (e.g. properties of cellular membranes) occurring during dialysis or as a result of dialysis.

Apart from its use in nephrology, the model could be potentially applied in research related to other medical disciplines, such as in emergency medicine for the analysis of the cardiovascular response to different kinds of systemic perturbations (e.g. a haemorrhage) or to study the effects of fluid infusions in fluid resuscitation (e.g. for the analysis of impact of the composition of the infusion fluid on the short- or long-term blood volume restoration).

References

1. Cavalcanti, S., Cavani, S., Santoro, A.: Role of short-term regulatory mechanisms on pressure response to hemodialysis-induced hypovolemia. Kidney Int. **61**(1), 228–238 (2002)
2. Cavani, S., Cavalcanti, S., Avanzolini, G.: Model based sensitivity analysis of arterial pressure response to hemodialysis induced hypovolemia. ASAIO J. **47**(4), 377–388 (2001)
3. Pstras, L., Thomaseth, K., Waniewski, J.: Personalised simulation of haemodynamic response to the Valsalva manoeuvre. In: Computational methods in data analysis, ITRIA 2015, Warsaw, pp. 119–134 (2015)
4. Pstras, L., Thomaseth, K., Waniewski, J., Balzani, I., Bellavere, F.: Mathematical modelling of cardiovascular response to the Valsalva manoeuvre. Math. Med. Biol. **34**(2), 261–292 (2017)
5. Schneditz, D., Roob, J., Oswald, M., Pogglitsch, H., Moser, M., Kenner, T., Binswanger, U.: Nature and rate of vascular refilling during hemodialysis and ultrafiltration. Kidney Int. **42**(6), 1425–1433 (1992)
6. Guyton, A., Coleman, T., Granger, H.: Circulation: overall regulation. Annu. Rev. Physiol. **34**, 13–46 (1972)
7. Coleman, T., Randall, J.: HUMAN. A comprehensive physiological model. Physiologist. **26**(1), 15–21 (1983)
8. Debowska, M., Poleszczuk, J., Dabrowski, W., Wojcik-Zaluska, A., Zaluska, W., Waniewski, J.: Impact of hemodialysis on cardiovascular system assessed by pulse wave analysis. PLoS One. **13**(11), e0206446 (2018)
9. Poleszczuk, J., Debowska, M., Dabrowski, W., Wojcik-Zaluska, A., Zaluska, W., Waniewski, J.: Patient-specific pulse wave propagation model identifies cardiovascular risk characteristics in hemodialysis patients. PLoS Comput. Biol. **14**(9), e1006417 (2018)
10. Schneditz, D., Van Stone, J., Daugirdas, J.: A regional blood circulation alternative to in-series two compartment urea kinetic modeling. ASAIO J. **39**(3), M573–M577 (1993)

Index

L. Petras, J. Waniewski, *Mathematical Modelling of Haemodialysis*,
https://doi.org/10.1007/978-3-030-21410-4

Printed in the United States
By Bookmasters